Surface Mount Technology for Concurrent Engineering and Manufacturing

Electronic Packaging and Interconnection Series
Charles Harper, Series Advisor

Published Books

HARPER • *Electronic Packaging and Interconnection Handbook*

HARPER AND MILLER • *Electronic Packaging, Microelectronics, and Interconnection Dictionary*

Related Books of Interest

BENSON • *Audio Engineering Handbook*

BENSON AND WHITAKER • *Television Engineering Handbook*

BOSWELL • *Subcontracting Electronics*

BOSWELL AND WICKAM • *Surface Mount Guidelines for Process Control, Quality, and Reliability*

BYERS • *Printed Circuit Board Design with Microcomputers*

CAPILLO • *Surface Mount Technology*

CHEN • *Computer Engineering Handbook*

COOMBS • *Printed Circuits Handbook*

DI GIACOMO • *Digital Bus Handbook*

DI GIACOMO • *VLSI Handbook*

FINK AND CHRISTIANSEN • *Electronics Engineers' Handbook*

GINSBERG • *Printed Circuit Board Design*

JOHNSON • *Antenna Engineering Handbook*

JURAN AND GRYNA • *Juran's Quality Control Handbook*

KAUFMAN AND SEIDMAN • *Handbook of Electronics Calculations*

LENK • *McGraw-Hill Electronic Testing Handbook*

LENK • *McGraw-Hill Circuit Encyclopedia and Troubleshooting Guide*

LENK • *Lenk's Digital Handbook*

PERRY • *VHDL*

RAO • *Multilevel Interconnection Technology*

SZE • *VLSI Technology*

TUMA • *Engineering Mathematics Handbook*

VAN ZANT • *Microchip Fabrication*

WAYNANT • *Electro-optics Handbook*

To order, or to receive additional information on these or any other McGraw-Hill titles, please call 1-800-822-8158 in the United States. In other countries, please contact your local McGraw-Hill office.

MH93

Surface Mount Technology for Concurrent Engineering and Manufacturing

Frank Classon

McGraw-Hill, Inc.

New York   San Francisco   Washington, D.C.   Auckland   Bogotá
Caracas   Lisbon   London   Madrid   Mexico City   Milan
Montreal   New Delhi   San Juan   Singapore
Sydney   Tokyo   Toronto

Library of Congress Cataloging-in-Publication Data

Classon, Frank.
    Surface mount technology for concurrent engineering and
  manufacturing / Frank Classon
        p.   cm. — (Electronic packaging and interconnection series)
    Includes index.
    ISBN 0-07-011200-2
    1. Electronic packaging.   2. Surface mount technology.
  3. Concurrent engineering.   I. Title.   II. Series.
  TK7870.15.C59   1993
  621.3815'31—dc20   92-4704
                                                                        CIP

    3 4 5 6 7 8 9 0   DOC/DOC   9 9 8 7 6 5 4

ISBN   0-07-011200-2

*The sponsoring editor for this book was Daniel A. Gonneau, the editing
supervisor was Caroline Levine, and the production supervisor was
Suzanne W. Babeuf. This book was set in Century Schoolbook by
McGraw-Hill's Professional Book Group composition unit.*

Printed and bound by R. R. Donnelley & Sons Company.

Contents

Contents

Preface

Integrating surface mount technology (SMT) into a company can have a jarring impact on the company's operating divisions if not done with care, caution, and some knowledge of SMT. SMT is so pervasive and exacting that all operating divisions within a company must approach it in the spirit of an interdivisional joint venture.

This book provides basic SMT design and manufacturing guidelines to equip concurrent engineering team members with the necessary basics and total overview to deal with the requirements driving companywide SMT activities. The book also gives management a quick overview of the scope and impact SMT can have on a company and how each division integrates with the whole.

Formatted for convenient reference yet detailed enough for implementation, the book goes far beyond the traditional detailed coverage of purely design issues and covers every aspect of SMT that a company should know. Its focus is on the integration of multiple functions within the company: design, manufacturing, quality, testing, procurement, material control, and management, with special emphasis for the individuals who have SMT decision-making and implementation responsibilities. It presents what every engineer should know about SMT and how to relate within the concurrent engineering environment.

It is assumed that the reader has some familiarity with electronic design and fabrication processes as practiced for insertion mounting technology. The predominant focus is kept on SMT throughout the book with few exceptions. Chapter 1 contains an introduction and overview on implementing SMT into the total company and how it relates to world-wide market forces within the electronics industry. Chapter 2 deals with technical fundamentals in design and manufacturing. Chapter 3 covers the full complement of SMT components available today. Chapter 4 is focused on the factory, the selection of SMT manufacturing machinery, and the integration of a computer-based manufacturing control system. Chapter 5 is dedicated to the how, why, and what of soldering. Chapter 6 presents quality and the impact that the new market-induced emphasis on quality upgrade has on a company. Chapter 7 deals with *electrostatic discharge* (ESD), a topic that has been deservedly overworked. ESD and SMT have been on a collision course from the beginning; to not emphasize ESD in a SMT treatise would be a mistake. Chapter 8 covers the SMT aspects of testing and leaves full coverage of testing for other volumes dedicated to this subject. Chapter 9 sets forth the current practices of SMT rework and repair. Chapter 10 presents the future trends in SMT and insights as to when and how some trends, already underway, will affect SMT. Appendixes A, B, C, and D provide ancillary topics that have some value for the reader.

Acknowledgments

This book is the beneficiary of the creative work of many of my colleagues at the Martin Marietta Corporation and the IPC in advancing the state of the art of surface mount technology. I am especially grateful for my association with my friend and mentor Rene Sandeau and my friend, the creative electronic wizard, Dave Emmons.

I wish also to thank those companies and individuals who took the time and effort to supply the photographs presented in this book.

Frank Classon

Implementing SMT into the Total Company

Surface mount technology (SMT) is a technology that has evolved as a natural progression from two predecessory technologies. It has evolved out of the insertion mounting technology (IMT) and the hybrid technology (see Fig. 1.1). There is nothing technologically new in SMT that was not first developed and perfected in the two earlier technologies. What is new in SMT is the way in which already developed materials, components, interconnection methods, mounting techniques, attachment processes, and assembly routines have been blended and exploited to form the new technology (see Fig. 1.2).

By taking subminiaturized passive chip components and microelectronic die chips from the hybrid technology and combining them with altered component features from IMT, the electronics industry has fashioned a new technology that has become known as SMT. It can

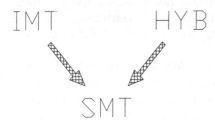

LEGEND :

IMT - INSERTION MOUNT TECHNOLOGY
HYB - HYBRID TECHNOLOGY
SMT - SURFACE MOUNT TECHNOLOGY

Figure 1.1 Evolution of SMT.

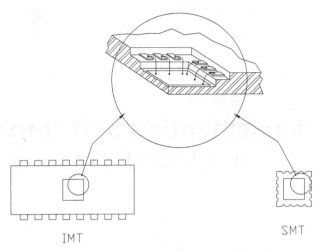

Figure 1.2   16-pin active device, IMT versus SMT.

safely be said, therefore, that SMT presents no new or risky technologies that are not already known and understood by industry. All the
basic elements that make up present-day SMT have been previously incorporated, in one way or another, in hybrid and IMT production products. Products now designed with the evolutionary SMT technology
are, in turn, being successfully produced in small and large quantities.

It should not be inferred, however, that SMT is, or was, problem-free.
Many difficulties unique to SMT have been encountered by industry
along the way, and all of them have been successfully dealt with and
resolved. Most problems encountered today associated with standard
size (1.27 mm) and fine-pitch size (0.64 mm) SMT lead pitch require not
necessarily reinvention but improvements to otherwise well-established designs and manufacturing procedures. Yet, there still is a need
in SMT for new inventions; new inventions not associated with established SMT procedures and processes, but new inventions associated
with efforts to push the state of the art toward a greater and greater
subminiaturization.

Estimates vary widely as to how much SMT has impacted the industry so far and how much it will impact it in the near future. Today's
worldwide use of one or more SMT components in standard products
has been estimated to be anywhere from 30 to 70 percent, and by the
mid-1990s electronic products will be dominated by SMT components.

Companies who are contemplating new products and who are not
currently using SMT are facing a dilemma. Should their new designs
be based on IMT, the technology which they know and which at present

can perhaps offer better component availability and even, in some cases, lower costs, or should they base their new design on a technology that is new to them and relatively new to the industry? A switch to the new SMT technology will require these companies to retool their factories and retrain their staffs. Facing this dilemma is not unique to any one company or any one market segment of industry. All companies in the electronics industry, both large and small, have or will have to face the SMT dilemma. No one can escape this dilemma and feel completely comfortable in today's changing market. It may be that continuing with IMT is the wise decision for a given company and a given product; but that decision can be properly made only after facing the SMT dilemma and going through SMT-versus-IMT benefits-consequences trade study.

1.1 SMT Benefits

Of all the many benefits inherent with SMT, one stands out above all the others: SMT makes it possible for industry to go far beyond the peaked configuration and production limitations of the existing IMT technology. IMT is a technology that has been fully exploited to its ultimate performance and has reached stagnation. SMT ushers in a myriad of new design and production solutions that make it possible to significantly subminiaturize electronic products and at the same time make them more reliable and cost-competitive, with added capabilities.

What SMT brings to the electronics industry is a new beginning, a next logical rung up the progression ladder of technological advancement.

Compared to traditional IMT, SMT presents a totally new concept for electronic packaging with wider design options and many more benefits. By eliminating the need for third-axis (Z-axis) mounting of electronic components, making it possible for them to be mounted in the X–Y plane only, and by miniaturizing component bodies and terminations, electronic products can be designed much smaller with less material and produced much more efficiently through total automation.

Two essential features are responsible for the majority of benefits derived from SMT:

1. Elimination of the Z axis in component mounting, i.e., mounting that is limited to the surface only.

2. The significant reduction in the sizes of component and terminations.

These two features account for the reduced size and weight of end products, increased electrical performance, improved reliability, cost sav-

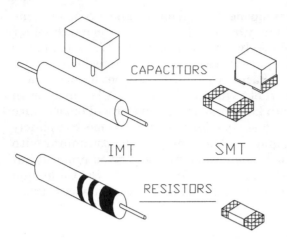

Figure 1.3 Passive devices, IMT versus SMT.

LEAD COUNT	RATIO OF AREAS	RATIO OF WEIGHTS
16–18	3 TO 1	10 TO 1
24–32	4 TO 1	12 TO 1
40–44	5 TO 1	14 TO 1
64–68	6 TO 1	7 TO 1

Figure 1.4 Size and weight, IMT versus SMT.

ings, and reduced factory size (see Figs. 1.3 and 1.4). The smaller component sizes, the reduced length of conductor runs, the elimination of component lead through holes, and the use of double-sided assemblies together make it possible, on a circuit-for-circuit basis, to produce SMT printed wiring board with smaller boards and fewer layers. SMT assembly machinery is smaller than its counterpart in IMT, requiring less factory floor space. Reliability is enhanced by the reduced component size, and production is automated.

SMT, with its small components, low profiles, and no-lead protrusions, has opened up extensive new vistas of electronic design possibilities. Because of these favorable component features it is now possible to nest four circuit card assemblies (CCAs) within a 2.54-mm stack, reduce the size of chassis assemblies from shoe boxes to cigarette packs, eliminate more than half the CCAs in most existing assemblies, and significantly reduce the number of system interconnections. All this reduction is achieved through SMT while simultaneously increasing the functions and reliability of electronic products.

Because of the smaller components and reduced assembly sizes, mechanical environmental survivability has also been significantly improved. Because of the low mass and low center of gravity of SMT components, products can be rendered more capable of surviving environments associated with vigorous handling, airborne installations, mobile ground vehicle mounting, and general outdoor life. With SMT, consumer electronics are much more able to go where humans go.

1.2   Future Trends in IMT and SMT

Most all new growth in the electronics industry has been accomplished using SMT. IMT continues to dominate the industry only because well-established, previously designed products remain in production. Even in a number of these products SMT components are now being used more. The prominence of IMT will, most likely, diminish before this decade ends.

1.2.1   Ongoing evolution

SMT is evolutionary and not revolutionary. The ever-increasing demand for faster circuits, greater subminiaturization, higher reliability, and reduced costs has forced the development of subtechnologies. The status of these newer subtechnologies is in a continuous state of flux. New inventions have been offered, and still many more are needed. New machines and new processes have been introduced. Wider use of basic technologies, such as infrared, laser, and X-ray; ultrasonics; optics; sensor scanning; image enhancement; and data processing are becoming essential features of production machinery used to fabricate and inspect products using these newer subtechnologies.

1.2.2   SMT subtechnologies

Identifiable subtechnologies have evolved out of the ongoing evolutionary activities in electronic packaging. These subtechnologies include

1. Fine-pitch components (FPs)

2. Multichip modules (MCMs)

3. Tape automatic bonding (TAB)

4. Chip-mounted-technology (CMT)—formerly chip-on-board (COB)

SMT can be categorized, now that these subtechnologies have been introduced, in accordance with the pitch of the components as follows:

1. Standard pitch—1.27 mm

2. Fine pitch—0.64 mm and lower

Multiples of silicon chips forming a functional circuit are being assembled into single-surface-mounted modules that are hermetically sealed. These multichip modules are then subsequently mounted to PWB assemblies. These multichip modules with their hermetically sealed packages and substrates with micrometer traces and spaces make it possible to mount bare silicon chips very close together, reducing substrate sizes and thereby reducing functional circuit time delays within the module. In addition to improving circuit speed within the module, MCMs also improve assembly miniaturization. The single-MCM package has, in effect, replaced multiple single-chip packages at a net space savings (see Fig. 1.5).

Until recently, microelectronic chips were interconnected within their individual packages by sequentially applied, discrete wires. A relatively new interconnection system that uses miniature flex circuits, referred to as *tape automatic bonding,* makes simultaneous termination possible. Tape automatic bonding is now being used to interconnect chips within some MCM packages.

Tape automatic bonding makes it possible to functionally test the bare chips prior to committing them to either single- or multichip packaging.

The next significant gain in miniaturization will be achieved when silicon chips, minus their packages, can be mounted directly on PWBs. Once silicon chips, and their interconnection terminations, are sealed with a long-term moisture barrier coat, sealed chips will replace pack-

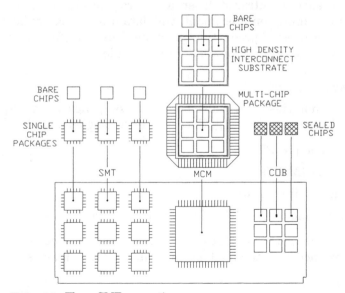

Figure 1.5 Three SMT generations.

aged chips. At that time production of individual ceramic packages and those MCMs which were created for miniaturization purposes will diminish. These sealed chips will be terminated with mass bondable tape automatic bonding.

Chip-mounted technology is presently being used for consumer products such as watches, calculators, and laptop computers. The moisture-sensitive, fragile chips are carefully bonded and terminated within these products in clean rooms and then coated with an encapsulation that protects them from moisture for a few years in relatively benign environments.

Industry has been steadily improving the moisture sealing duration of chip encapsulants. Present estimates are that an adequate military-grade encapsulant will be developed before the mid-1990s. The telecommunications and automotive industries have begun using silicone gels to seal chips in their products.

Industry will best be served, once the chip encapsulant has been perfected, when chips are inner-lead-bonded to individual lead patterns serially placed on a continuous reel of tape, tested, encapsulated, and shipped to the user while still on the tape reel. The user would excise the terminated and encapsulated chip from the tape reel and mount it to the PWB as an ordinary component having ultra-fine-pitch leads.

Significant miniaturization gains will be achieved when a successful chip sealing coat is proved and acceptable (see Fig. 1.5).

1.3 Entering into SMT Production

SMT can easily be fully automated: the equipment and the knowledge are readily available. With full automation, product quality becomes nearly totally dependent on the quality of the machines, material, and processes, and the manufacturability of the design. Large-scale production of SMT products are therefore generally not the output of individuals or isolated groups but the output of an integrated team, integrated machines, and integrated computers. There is, consequentially, an unavoidable interdivisional interdependence between all of a company's operating divisions.

Unlike earlier technologies, where a company's operating divisions, such as engineering and manufacturing, could successfully perform their individual functions somewhat independently from one another, SMT requires a start-to-finish involvement of those two divisions along with all other company divisions involved in production. Almost all decisions and actions taken by individual divisions need to abide in mutual consent, knowledge, and cooperation between all other divisions.

It is essential that all interdivisional personnel from all divisions have a working knowledge, to one level or another, of SMT requirements and the extent of contributions made by other divisions. In SMT, production control processes have as much impact on product liability as does design; procurement is as much involved with the earlier stages of product planning and matériel decisions as is the manufacturing engineer; the contributions made by the quality staff are as vital as those made by management. The first step for a company, after deciding to implement SMT, is to form an SMT team with representatives from engineering, manufacturing, quality, procurement, material control, facilities, finance, personnel, and management. SMT can also be implemented with less than full automation.

Factories can be sized to match the production throughput rate needs of almost all companies. Machines are available to efficiently produce quantities of SMT assemblies from low to medium to high production. Production and inspection machine costs start at $2,000 and exceed $1,000,000.

Facilities, skills, operating routines, schedules, inventory, design, budgets, and suppliers will all radically change with the introduction of SMT. The requirement for total company cooperation and involvement from the beginning cannot be emphasized enough; without it, implementation of SMT will fail.

Although this book focuses primarily on the fundamentals of SMT design and manufacturing, it has a strong secondary aim focused on management and control and procurement.

The question for most electronics companies is no longer "Should we include SMT in our new products?" or "How soon should we implement SMT?" but "How quickly can we convert?" SMT is rapidly becoming the technology of choice for most all new product designs in most all segments of the electronics industry.

Bibliography

Allen, Kevin, and Anna Hargis: "Surface Mount Assembly: Meeting the Challenge," *Circuit Assembly,* September 1991, pp. 30–39.

Bindra, Ashok: "SMT Changes PC-Board Industry," *Electronic Engineering Times,* April 17, 1989, pp. 117, 122.

Davis, Dwight B.: "Rosy Decade Looms for Communications Market," *Electronic Business,* December 11, 1989, p. 34.

Derman, Glenda: "ICs Count on SMT Packages," *Electronic Engineering Times,* February 25, 1991, pp. 66, 90, 91, 108.

DeYoung, Garrett: "Global Challenges and Promises in the 1990's," *Electronic Business,* December 11, 1989, pp. 10–13.

Huey, Steven: "Management Responds to Surface Mount's Growing Impact," October 1989, pp. 88, 89.

"Implementing SMT: A Painful Rite of Passage for Board Assemblers," *Electronic Packaging and Production,* October 1989, pp. 108–110.

Kuhn, Harry: "For Military, It's SMT," *Electronic Engineering Times,* July 10, 1989, pp. T12, T16.

Lapin, Philip J.: "Internationalization," *PC FAB (Printed Circuit Fabrication)*, February 1992, pp. 108–110.

Leibowitz, Michael R.: "Chips Will Soon Take Over the World," *Electronic Business,* December 11, 1989, p. 33.

Lilja, Dean: "Smoothing the Transition from Through-Hole to Surface Mount Design," *Surface Mount Technology,* February 1991, pp. 56–58.

Mangin, Charles-Henri: "SMT Snapshot: U.S. Electronics Assembly Logs," *Electronic Business,* June 15, 1988, p. 9.

Nash, Thomas F.: "Mixed Technology Production: The Real Revolution," August 1991, p. 90.

Sandeau, Rene F.: "Solution to the SMT Maze," *TEPS Journal,* vol. 9, no. 3, pp. 14–16.

Staff Writer: "500 Percent Growth in Automotive Electronics Seen by 2000," *Electronics Manufacturing,* August 1990, p. 10.

Stout, Gail: "The Military: A Bright Future for Surface Mount," *Surface Mount Technology,* April 1989, p. 4.

———, "SMT Market Trends," *Surface Mount Technology,* November 1990, p. 4.

SMT Design Fundamentals

2.1 Scope of SMT Design

The scope of design possibilities for the electronic packaging engineer has broadened considerably with the introduction of SMT. With SMT it is possible to take full advantage of the electronic miniaturization gains achieved by the semiconductor industry and, with those gains, make sizable reductions in electronic products with increased functional capability in reliability and reduction in system costs.

2.2 Design Issues

Many of the design guidelines and standards developed for IMT are still valid for SMT. There are, however, several design issues unique to SMT for which some guidelines and standards have been established and others that are still in development. The most important SMT design issues are

1. SMT assembly types

2. SMT product design done concurrently with manufacturing and others

3. Coefficient of thermal expansion (CTE) mismatch between different materials used for SMT PWBs and assemblies

4. Thermal management of higher heat densities

5. CAD and CAE (computer-assisted design and engineering) in SMT design

6. Packaging density and component footprint pattern design limits

7. Vias and PTHs

8. SMT reliability

9. High-yield, first-pass manufacturability

10. Material selection

11. Designing for test

2.2.1 SMT assembly types

Three general types of SMT circuit card assemblies (CCAs) are prevalent during this evolutionary period from IMT to SMT. Each type is categorized according to the soldering operation(s) associated with it. Assemblies that require reflow soldering only are categorized as Type I. Accordingly, this type comprises only SMT components mounted on either one or both sides of the PWB. Assemblies that require a combination of reflow soldering and wave soldering are classified as Type II. These assemblies have a mix of SMT and IMT components with IMT only on the top side and SMT on one or both sides. The third category, classified as Type III, is reserved for wave soldering only and consists of IMT components on the top side and SMT components on the second, wave-soldered side (see Fig. 2.1).

Type I is used when the maximum SMT miniaturization benefits are required, or desired, and all the active and passive components are available in SMT configurations. Type II is used when some SMT components are not available and miniaturization is still important. Category Type III is used when size, weight, and electrical characteristics are not the dominant design requirements and when active SMT components are unavailable and/or costs of a total SMT assembly are

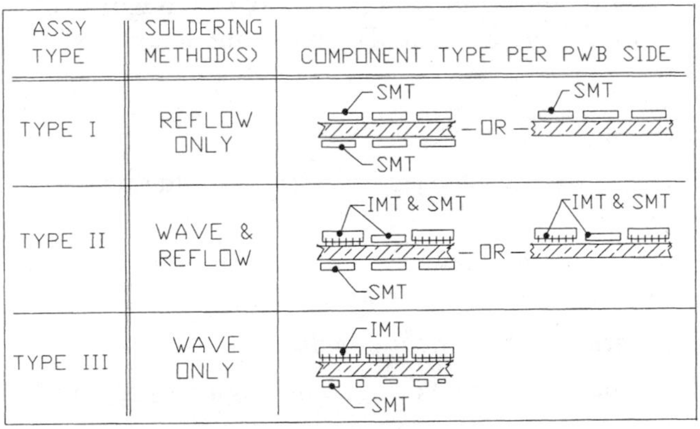

ASSY TYPE	SOLDERING METHOD(S)	COMPONENT TYPE PER PWB SIDE
TYPE I	REFLOW ONLY	SMT / SMT — OR — SMT / SMT
TYPE II	WAVE & REFLOW	IMT & SMT / SMT — OR — IMT & SMT
TYPE III	WAVE ONLY	IMT / SMT

Figure 2.1   SMT assembly types.

ASSY TYPE	ADVANTAGES				DISADVANTAGES		COMMENTS
	1	2	3	4	5	6	
I	MAX DENSITY	ONE SOLDER METHOD	LESS MFG FLOOR	ONE ASSY MACHINE	COMP. AVAIL.	LEARN NEW METHODS	HIGHEST DENSITY
II	COMP. AVAIL./ COST	HIGH DENSITY TECH'	–	–	DUAL SOLDER TECH'.	COMPLEX PROCESS	SMT/IMT ON SAME SIDE
III	FAMILI- ARITY	NO NEW SOLDER	–	–	SMT SOLDER IMMERS	NON- DENCE SMT	SMT GLUED ON SIDE TWO

Figure 2.2 Assembly types—pros and cons.

unacceptable. Type III is sometimes referred to as the "transitional design." This type of assembly is often used by companies just getting started into SMT.

Designing CCAs on a circuit-by-circuit basis is less difficult and less complex with SMT than with IMT. SMT CCA designs, however, are rarely direct circuit-by-circuit conversions from IMT but are more often circuits that transcend the capabilities of IMT and push the state of the art in terms of miniaturization, functionality, and reliability.

Designing the ideal SMT assembly involves the pursuance of an arduous series of interrelated design and manufacturing trade studies and trades made between traditional materials and state-of-the-art materials, between component types, between fabrication techniques, processes, and between circuit partitioning and tests (see Fig. 2.2).

2.2.2 Concurrent engineering

New marketing forces are driving established electronics companies to new operating procedures. To stay competitive, companies must now significantly reduce the time-to-market for new products and do so at less cost. Previously, new product manufacturing decisions were postponed until the design was established. Now, in the interests of shorter turnaround time and the need for a single-pass manufacturing cycle, new product manufacturing decisions must be made up front and concurrently with design decisions. For design engineers, priority weighing of design decisions can no longer be made in favor of functional factors at the expense of producibility factors. The design decision priority for producibility must now be elevated to a status equal to that of form, fit, and function. For time-to-market and single pass to work, decision-making personnel from the production divisions must now participate with the design engineer in establishing new product designs that balance function with producibility.

Design versus factory capabilities. There has always been a direct link between engineering designs and factory capabilities, and in the past engineering has often been the sole arbitrator on how that was reflected in the design. With SMT there is a cost imperative for multidivisional participation in new product designs. Capital expenditures for SMT machinery are much higher than for its counterpart IMT machinery. Once a mass-production factory has been committed to a particular set of SMT production machines, the amount and type of design configuration latitudes for new products have been somewhat predetermined and limited. It is simply too expensive to belatedly alter mass-production machines to conform to individual design features. There is, however, nearly always a marketing conflict between the particular configuration of a new product and the need to conform to company manufacturing capabilities. In this new era of global competition, companies can no longer afford to have their design engineers solely resolve the conflict between design, manufacturing, and market forces. How design trade studies are conducted and what conclusions are reached has a direct impact on market response and profits.

Start of interdivisional activity. Involvement by decision-making personnel from all operating divisions within the company is required at the very start of an SMT project and, together with engineering, establishes the company's design and production strategy. This body of people should serve both macro and micro SMT functions. They are vital contributors to the difficult and crucial task of (1) balancing the marketing needs against the production, (2) setting design configuration limits, (3) determining the use of contract manufacturers, (4) determining inventory, (5) arranging material flow, (6) negotiating supplier affiliations, (7) ascertaining capital expenditures, and (8) resolving customer commitments. All of these issues are properly considered by this interdivisional team. This body of people also serves to check and double-check the design details and process development details as these details affect each of their particular production-related functions. The goal and commitment for the company, all operating divisions, and this team, should be a "one-time pass" through the production cycle and test phases.

When the concurrent engineering process is successful, manufacturing machinery and tools are in position and free to be used for the new project, components are packaged in kits and available on the assembly line, people are trained, fabrication processes are developed, test equipment is proved, software programs are certified, and then the assembly-line start button can be pushed as the ink is drying from the last signature entered on the engineering drawings. This design in-

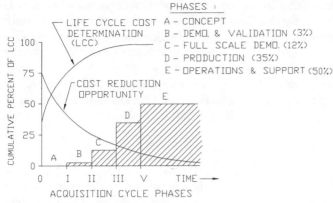

Figure 2.3   Life-cycle cost determinate versus time.

volvement by the total company applies to prototype as well as production projects. In fact, it is in the prototype phase where the greatest downstream production payback is most often determined and achieved (see Fig. 2.3).

Multifunctional SMT team.   Companies who have been successful in establishing total company involvement in the SMT design process from concept onward have all done so by forming a permanent SMT multifunctional team consisting of key members representing each of the company operational divisions: engineering, manufacturing, quality control, purchasing, production control, marketing, accounting, tests, staffing, and training. The team is generally charged with three levels of responsibility. The first level involves making overall determination of the company's SMT goals, setting milestones, making plans, initiating action, monitoring processes, and correcting the system as necessary.

The second level of responsibility is the establishment of company guidelines and standards. Here, industry experience and publications play a major role in getting started, but as company experience is gained, these earlier industry-based company documents are generally altered to fit the individual company's idiosyncrasies and incorporate "lessons learned" from each project. The third level is involved with specific design projects. This can be done by a second, smaller team, or by the original team. This team becomes actively involved from the very beginning in determining specific design strategies for any given project and stays with the design process all the way through to final drawing releases. Engineering does the actual design; the rest of the team serves as participating consultants and as representatives from the various divisions who influence design particulars and who can effect early action taken within their divisions in preparation for a timely turn-on of the production process as the design comes to fruition. The team leader is often, but not necessarily, the engineering representative or the manufacturing representative.

The requirement for total company involvement and cooperation from the beginning of the design phase cannot be emphasized enough; without it, implementation of SMT will fail.

2.2.3 Coefficient of thermal expansion

Materials expand or contract isotopically when subjected to temperature variations. Each type of material has its own characteristic rate of change per temperature change; a characteristic, referred to as the *coefficient of thermal expansion* (CTE) and measured in parts per million per degree Celsius (ppm/°C). In SMT, coping with this thermal characteristic is a major design consideration when selecting materials and component lead configurations.

When different materials that are fixed in intimate contact within an assembly or laminate are subjected to temperature variations, each material changes size in response to its own individual CTE characteristic. As the temperature change increases, an ensuing physical conflict between the materials develops, resulting in the generation of stress forces that put the materials under strain and, with time and recurrence, fatigue the materials.

In general, coping with CTE mismatch in most mechanical designs is important; coping with CTE mismatch in SMT is vital. Success or failure of SMT designs can be determined by how well the design takes into account the CTE mismatch between the components, solder, and the PWB materials. The CTE for leadless ceramic chip carriers (LLCCs), for example, is approximately 6 ppm and the CTE for FR4 laminate is approximately 17 ppm. The 11-ppm mismatch between the two is a minor problem for small-sized components used in benign environments, but this amount of mismatch would be a major problem for large components used in nonbenign environments. The goal in SMT design is to minimize the amount of CTE mismatch stresses on solder joints. There are several ways to minimize solder joint stresses that involve choice of components and choice of materials. The following four methods are the dominant techniques currently used:

1. Reduce the distance between two diagonal corner solder joints on LLCCs by limiting the number of component leads.

2. Use leaded components with compliant leads in lieu of leadless components.

3. Constrain the amount of CTE-induced movement in the X-Y plane of the PWB by using reinforcing fibers containing ultralow CTE valves.

4. Use sheet stock material with low CTE values either as embedded laminate layers or as externally applied plates structurally bonded to the PWB to constrain the PWB CTE.

In most SMT designs the CTE mismatch solution involves a combination of two or more of the above methods.

Methods 1 and 2–Compliant link.   Each material type has its own CTE value, and when two or more materials are fixed together in a laminate or assembly, there must be a compliant link between them to compensate for the thermally induced dimensional mismatch; otherwise damage will ensue. When LLCC components are soldered to PWBs, the solder joint serves as the sole compliant link. Solder is a capable compliant link, but it has limited elasticity and limited fatigue life and therefore, when used as the sole compliant link, it must be done prudently.

In method 1 solder continues to be successfully used as the compliant link compensating the mismatch between PWBs and most all passive chip components, metal-electrode faced (components) (MELFs), small-outline transistors (SOTs), small-outline integrated circuits (SOICs), and LLCCs with 28 leads or less. Solder is also successfully used for larger LLCCs when the operational environment is relatively benign.

One alternative to using solder as the compliant link is to use method 2 components having compliant leads. Compliant leads on components are more able than solder alone to cope with the CTE mismatch stresses encountered in solder joints on components having more than 28 leads that are also used in nonbenign environments. Compliant component leads also have a much longer fatigue life than does the solder. For these reasons, leaded ceramic chip carriers (LDCCs) in hermetically sealed applications and leaded plastic carriers in nonhermetical applications are used to replace large LLCCs in the nonbenign environments.

LLCCs are much less costly than LDCCs, however, and for this reason LLCCs are used in lieu of LDCCs whenever the environment permits or the PWB CTE is constrained (see Table 2.1).

Methods 3 and 4–Modified PWB material to lower CTE.   Larger-sized, standard pitched (1.27-mm) LLCC devices can be used for the more aggressive nonbenign environments by lowering the CTE of the PWB in the X-Y plane. In method 3, the reinforcing glass fibers normally

TABLE 2.1   LLCC and Material Combination—1000-Cycle Survival

Temperature range, °C[b]	PWB materials	PWBT_g, °C	Heat sink	Adhesive type	Maximum-size LLCC by quantity of leads[a]
0 to +70[a]	Epoxy glass	125	—	—	28
0 to +70[b]	Polyimide glass	230	—	—	28
−30 to +70[c]	Tetrafunctional epoxy glass	135	—	—	—[g]
−50 to +100[d]	Polyimide glass	230	CIC	Rigid	68
−50 to +100[e]	Tetrafunctional epoxy aramide	135	Aluminum	Soft	44
−50 to +100[f]	Modified epoxy quartz	180	Aluminum	Soft	44
−55 to +100[g]	Polyimide flexible circuit	—	—	—	—[g]

[a]SOTs and leaded active components can be used with all listed materials and temperature ranges.

[b]1000-cycle survival and temperature ranges.

[c]Rigid adhesives form hard bonds between PWBs and heat sinks.

[d]Soft adhesives form compliant bonds between PWBs and heat sinks.

[e]All components have 1.27-mm lead pitch.

[f]Design approval required for military designs.

[g]SOTs and leaded active devices only.

used in conventional laminates can be replaced with a newer fiber having an ultralow CTE and a higher elastic modulus, thereby lowering the laminate CTE and the PWB CTE, thus reducing the dimensional mismatch between the ceramic component and the PWB in the X-Y plane. A second technique in method 3 uses conventional PWB glass fibers with a newer resin that has a CTE lower than that of conventional resins. Method 4 has two options. Option 1 uses constraining layers of copper-invar-copper (CIC) or copper molybdenum-copper (CMC) laminated into the multilayer board. These laminated layers constrain the PWB CTE in the X-Y plane. Option 2 uses thicker sheets (0.80 to 1.27 mm) of CIC or CMC rigidly bonded between two single-sided Type I SMT assemblies, thereby constraining the X-Y movement of both assemblies.

Method 4 is unique in that it serves four functional design roles simultaneously. The metal plane, in addition to constraining the CTE, serves as a heat dispenser, as a board stiffener, and as power and ground planes. Because of its multiple functions, method 4 has been extensively used for military assemblies. There are, however, several disadvantages to using method 4. The iron and copper contents of the constraining sheets make the boards heavier. By constraining the X-Y plane, the normal three-axis isotopical expansion of the laminate material is distorted. The laminate material compensates for being abnormally constrained in the X-Y plane. The greater the CTE constraint in the X-Y plane, the more the material is compelled to expand a greater amount in the Z axis. The stresses generated by this abnormally high expansion of the Z axis in the constrained SMT boards causes plated through-holes (PTHs), fabricated to traditional standards, to experience barrel cracking failures (see Sec. 2.2.8).

Strengthened PTHs. To structurally compensate the PTHs against the increased forces in the Z axis, the nominal PTH wall thicknesses are doubled from 0.025 to 0.050 mm (0.036 mm minimum), and, in some applications where the boards are thicker or the reliability requirements are higher, a layer of nickel plating (8 μm minimum) is also added. The minimum-size PTH through-hole also needs to be greater than 0.30-mm diameter for boards at least 1.27 mm thick.

Soldering operations are enhanced when nickel plating is employed. Solder will not wick into the nickel-plated holes that are attached to component pads. Selective nickel plating on traces can also prevent wicking away from FP pads. Copper-flashed layers can be applied over the nickel to enhance selective solder adhesion.

A fifth method and a sixth method of controlling CTE could also be added as design candidates; however, these methods have not yet been tried in production. The first of these two candidates is to use leadless plastic carriers. This method could be very cost-effective. Because it is organic, the component CTE could match the PWB CTE and eliminate the mismatch problem. In addition to the CTE benefit, the ease of handling and assembly attachment associated with leadless devices could also be gained. It may be possible, under these conditions, to use leadless plastic components for devices above 100 leads. With the use of leadless plastic carriers, reliability gains, in addition to cost gains, could also be achieved because the CTE of both the component and the PWB would more closely match the CTE of solder (24 ppm). Leadless plastic carriers, however, are not available at this time.

The sixth candidate would use a compliant top coat layer, either laminated or bonded to the PWB, to serve as the compliant link between conductor pads and component leads. Nitrile rubber, acrylic elastomers, nonreinforced polyamide, and other materials have been tried with varying success. Low temperature sensitivity, drill smear, and freedom of compliant movement inhibited by PTHs anchored to the noncompliant parent laminate are some of the obstacles to industry's interest in this alternative method.

2.2.4   Thermal management

Heat density. Heat density is the dominant thermal problem in SMT. Compared to active IMT components, active SMT components have smaller body sizes from which to dissipate comparable amounts of heat. Smaller body sizes allow tighter packaging densities, consequently resulting in higher heat densities, which complicates heat dissipation.

Heat dissipation is the dominant design aspect of SMT thermal management. As sizes continue to diminish and power continues to rise, heat density often increases beyond the cooling capability of passive systems. Semiactive and fully active cooling systems—forced air, heat pipes, thermoelectric, liquid cooling, and immersion cooling—are becoming more commonplace.

Component thermal pads.   Thermal transfer from the chip (die) junction through the component body into the substrate (PWB) can be aided by the addition of thermal pads on the bottom surface of the component and/or by thermal conductive adhesive, or solder columns, attaching the component thermal pads to the substrate (see Figs. 2.4 and 2.5).

Thermal vias.   Thermal transfer through the PWB into the heat sink can be greatly aided by the addition of extra PTHs in the PWB located

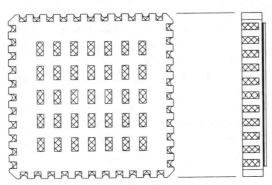

Figure 2.4   LLCC thermal pad array.

PKG	PIN QTY	MTG TYPE	MOLD COMPOUND	JA °C/W	JC °C/W
SO	14	SOCKET	TYPE A	113	29
SO	14	BOARD *	TYPE A	87	28
✓ SO	14	BOARD	TYPE A	121	30
SO	16	SOCKET	TYPE B	122	41
SO	16	BOARD *	TYPE B	105	40
SO	20	BOARD	TYPE A	97	27
SO	20	BOARD	TYPE B	109	38
PLCC	20	SOCKET	TYPE A	107	28
PLCC	20	BOARD *	TYPE A	72	26
PLCC	28	SOCKET	TYPE A	83	16
PLCC	28	BOARD *	TYPE A	52	16

NOTES

TYPE A MOLD COMPOUND Al_2O_3 FILLER. BEST THERMAL CAPABILITY
TYPE B MOLD COMPOUND: FUSED SILICA FILLER. LOWEST EXPANSION.
* PWB WITH COPPER BACK PLANE

Figure 2.5   SMT component thermal characteristics.

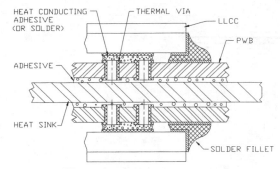

Figure 2.6 Thermal vias.

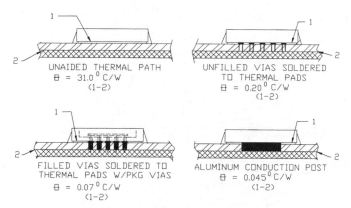

Figure 2.7 Component and PWB thermal resistance.

in alignment with the component thermal pads (see Figs. 2.6 and 2.7). No more than five or six vias are necessary for any one component.

Heat sink for dual CCAs. The addition of inner-core PWB layers of clad metals (CIC or CMC) or the attachment of a heat-sink plate between two back-to-back, single-sided CCAs improves heat transfer laterally through the PWB and into wedge-locked chassis walls (see Fig. 2.8).

Heat pipes. Dual CCA heat sinks consist mostly of flat metal plates made of aluminum, copper, clad metals, or composites. Heat pipes and hollow-core laminated metals, for air or liquid, are also used as heat sinks for dual CCAs (see Figs. 2.9 and 2.10).

Component location patterns for optimum heat transfer. Components that dissipate the most heat can be located relative to one another, and the edge of the CCA, to permit optimum heat transfer (see Figs. 2.11 and 2.12).

Heat sinks for individual components. Thermal grease, mica, thermal conductive pads, or thermal conductive adhesives are used beneath hot components to supplant the thermally poor condition through dead-air

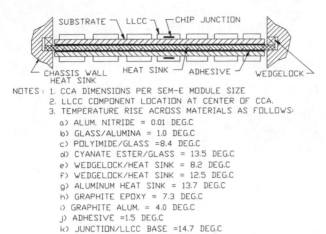

NOTES : 1. CCA DIMENSIONS PER SEM-E MODULE SIZE
2. LLCC COMPONENT LOCATION AT CENTER OF CCA.
3. TEMPERATURE RISE ACROSS MATERIALS AS FOLLOWS:
    a) ALUM. NITRIDE = 0.01 DEG.C
    b) GLASS/ALUMINA = 1.0 DEG.C
    c) POLYIMIDE/GLASS =8.4 DEG.C
    d) CYANATE ESTER/GLASS = 13.5 DEG.C
    e) WEDGELOCK/HEAT SINK = 8.2 DEG.C
    f) WEDGELOCK/HEAT SINK = 12.5 DEG.C
    g) ALUMINUM HEAT SINK = 13.7 DEG.C
    h) GRAPHITE EPOXY = 7.3 DEG.C
    i) GRAPHITE ALUM. = 4.0 DEG.C
    j) ADHESIVE =1.5 DEG.C
    k) JUNCTION/LLCC BASE =14.7 DEG.C

Figure 2.8 Temperature drop—junction through wedgelock.

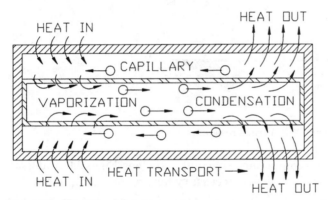

Figure 2.9 Heat pipe (plate).

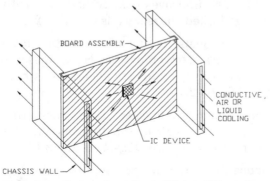

Figure 2.10 System heat flow—IC to ultimate heat sink.

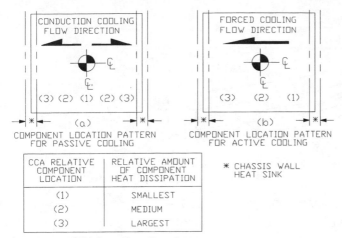

Figure 2.11   CCA heat flow direction.

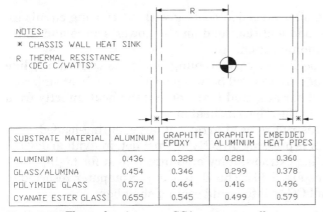

Figure 2.12   Thermal resistance—CCA center to wall.

pockets beneath individual components with a far better thermal conduction medium. When conduction and natural convection can no longer provide the necessary cooling, forced air may be necessary, provided there is available system volume for fans and ducts and the amount of heat to be carried off is within the range of forced-air cooling (see Fig. 2.13). A large variety of individual component heat sinks with monodirectional and bidirectional fins are on the market. These heat sinks are available as bond-ons, precision fits, and labor-saving clip-ons. Omnidirectional, fin-pin configurations are also available.

Thermoelectric heat pumps are ideal for high-thermal-density spot cooling associated with some SMT assembly components. These cooling

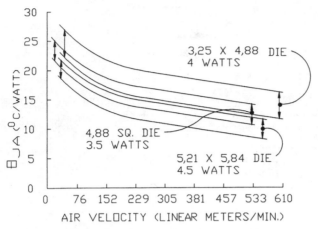

Figure 2.13 Thermal resistance—junction to ambient versus air velocity chip cooling.

units are small and offer precision cooling without causing circuit interference. They do add addition load on the power source and influence the heat-sink configuration.

Bags filled with perfluorocarbon cooling liquid function much like heat pipes. These bags are wedged between a CCA and a heat sink conforming to contoured surfaces and transferring the heat directly from the component case to the adjacent heat sink.

The computer industry has used arrays of large, spring-loaded thermal conductor plungers pressing directly against hot, individual devices as a way of dissipating extraordinary amounts of heat for high-performance machines (see Fig. 2.14). The Cray-2 supercomputer resorted to totally immersing all CCAs into liquid cooling chambers.

System cooling.   Military airborne radars and IR (infrared) sensor systems use liquid cooling through heat exchangers for portions of the system circuits.

Cryogenic cooling with liquid nitrogen, argon, or helium is reserved for rapid cooling of those microelectronic, SMT, sensor devices not nor-

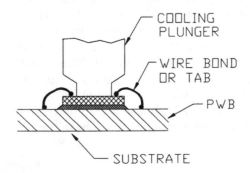

Figure 2.14   Chip cooling.

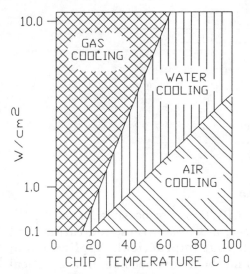

Figure 2.15 Chip cooling efficiencies.

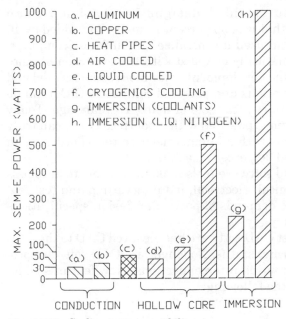

Figure 2.16 Cooling types—capability.

mally mounted to PWB assemblies. Cryogenic cooling takes chip temperatures down to the 80-K region for optimum function. CTE mismatch becomes acute at these temperatures, as does thermal shock.

Cooling capacities of various materials and media are shown in Figs. 2.15 and 2.16. Thermal environments per industrial segments are shown in Fig. 2.17.

ENVIRONMENT	INDUSTRIAL SEGMENTS			
	AUTOMOTIVE	COMMERCIAL	CONSUMER	MILITARY
TEMPERATURE	-40 TO 125^{0}C	0 TO 70^{0}C	0 TO 40^{0}C	-55 TO 125^{0}C
HUMILITY	85% RH/85^{0}C	CONTROLLED	NORMAL AMB.	85% RH/85^{0}C
SHOCK	150g	MINIMAL	MINIMAL	1K-100Kg
VIBRATION	20g,20-2K HZ	MINIMAL	MINIMAL	20-100g/ 20-20K HZ
CHEMICAL RESISTANCE	SALT SPRAY, AUTO FLUIDS	GENERALLY, NOT RESISTANCE	NOT RESISTANCE	SALT SPRAY,

Figure 2.17 Environments per industry segment.

2.2.5 CAD and CAE in SMT design

Manual layout. Manual layout of PWBs is not possible for most SMT designs. High-density tolerances, multiple layers, and turnaround design times make manual layout impractical. Application of computer-aided design (CAD) methods has become, in a practical sense, mandatory for all SMT PWB designs.

CAD. Initially, computer-aided designs involved a human (operator)/machine ratio that was 90 percent human effort aided by 10 percent machine effort and used a centralized, mainframe computer. That ratio has been dramatically changed with the evolution of more powerful computers and the development of far more sophisticated and complex software. Workstations and personal computers (PCs) can do today what mainframes were incapable of doing 10 years ago. Today mainframes and workstations powered with modern software can truly produce automated designs with a human/machine ratio bordering on the 10 percent human and 90 percent machine.

New CAD computer and software tools make it possible to automatically incorporate mechanical, electrical, manufacturing, and test features tailored to meet the demands of particular design specifications as well as those of the company.

Capabilities of the latest CAD software has elevated CAD to the CAE (computer-aided engineering) status. No longer are CAD software programs limited to simply routing interconnections. New features of design automation include the following:

1. Schematic capture

2. Timing simulators

3. Circuit simulators

4. Netlist testing

5. High-speed circuit routing

6. Component placement

7. Optimizing router algorithms

8. Multiple square and rectangular grids

9. Mechanical drawing interface

10. Thermal simulators

11. Automated data conversion to enhance data transfer interface to multiple machine types

12. Blind and buried via implementation

Finite element analysis CAE programs with wireframe models have been upgraded to solid models. Design tradeoffs can be analyzed to determine the thermal, mechanical, and electrical effects on the assembly when changing the locations of components and rerouting the conductors, or when exchanging heat sinks. These finite element models can be created directly from CAD data.

2.2.6   SMT packaging density and footprint patterns

Component density has a direct influence on SMT miniaturization, circuit speed, thermal control, weight, and manufacturability. In general, the higher the quantity of components in a given CCA area, the faster the circuit speed and the smaller the assembly size. However, with these advantages, brought about by higher densities, comes the disadvantages of higher heat densities and lower manufacturability. The critical and central issue in design is determining the concept that ideally balances size, weight, speed, thermal management, and manufacturing yields. Component types, PWB features, thermal control ingredients, and manufacturing processes are the key elements to be properly balanced to achieve cost objectives and customer satisfaction.

Density factor.   One of the most significant aspects of design planning, for SMT assemblies, is determining the packaging "density factor" (DF): the amount of circuitry that can be placed within a given amount of PWB area. The use of a DF helps conceptual design planning in four very significant ways. First, it makes it possible to reasonably partition the circuitry into individual PWBs. Second, it gives an indication of the complexity of assembly and PWB complexities, yields, and costs of the finished product. Third, design-manufacturability trade studies can be conducted that trade complexity against quantity of boards and then trade the quantity of boards against the size and cost of the final product. Finally, the DF provides a way of measuring design proficiency to determine whether the optimum layout has been reached.

Packaging density factors are the ratios of the board areas occupied by components compared to the board areas allocated for component placement. It is a single number that can readily communicate the complexity, and hence the manufacturability, of an SMT assembly and its PWB.

Component areas, used in the DF algorithm, consist of items 1 and 2 in the following list for passive devices and items 1, 2, and 3 for active devices:

1. The maximum dimensions of the component body shadow area

2. Marginal clearances around each component needed for placement apparatus, soldering devices, i.e., soldering bars and solder perform attachment fixtures, and/or fiducial symbols for components with lead pitches less than 1.00 mm

3. The peripheral area around the active components needed for mounting pads, vias, and test pads (see Fig. 2.18).

Component placement areas consist of the total, edge-to-edge, board area minus areas devoted to I/O connector(s), tooling features, and board mounting features.

The influence of passive devices on product size is deceptive. Every component, regardless of category, has an aura of space prescribed to it for normal overhead function, specifically, signal routing, placement tools, and repairs. The ratio of component size to allocated overhead space is significantly higher for smaller components. This ratio plays an important role in product sizing when the proportion of passive-to-active devices is larger than 2 to 1.

The amount of overhead area for passive devices ranges from 70 percent for the smaller devices down to 20 percent for the largest devices. Overhead areas for active devices range from 50 percent for 16-pin LLCCs down to 25 percent for 68-pin devices. Overhead areas for 68-lead, standard-pitch LDCC, and for 132-lead, 0.63-pitch devices are 45 percent.

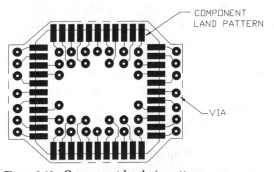

Figure 2.18 Component land-via pattern.

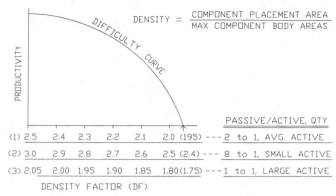

Figure 2.19 Packaging density versus productivity.

Producibility is directly related to DF, which is, in turn, related to the passive-to-active component mix.

As shown in Fig. 2.19, a graph depicting the producibility curve for an assembly with a 2.1 quantity mix between passive and active components, producibility decreases as the density factor decreases below 2.5, and drops below economic acceptability for 1.95. Figure 2.20 is another way of showing the relationship between the DF and passive-to-active component mix.

The DF numbers in Fig. 2.19 change for assemblies in which the predominant amount of components are the larger LLCCs and the remainder of components are a few passive devices. These DF numbers change again for assemblies with an 8.1 ratio of passive components to active components (see Fig. 2.20).

PWB features. Low-density PWBs have the highest manufacturing yields. Single-sided, double-sided, and simple multilayer boards (four layers) are in this category. PWBs in this category have 0.25-mm trace widths and spaces, 1.00-mm-diameter vias, 1.65-mm pads with 1.27-mm component lead pitch, and 0.63 × 2.03-mm pad sizes. Manufacturing yields are slightly decreased when the trace width and space is reduced to 0.20 mm and the lead pad to 0.63 × 1.90 mm and/or six or eight layers. Yields are significantly reduced when trace widths and spacings drop down to 0.13 mm, vias to 0.35 mm, and pads to 0.51 × 1.78 mm and/or 12 or more layers.

Component footprints. Footprint pad sizes and relative locations have a profound influence on the reliability of the final solder joint and factory yields. Arriving at one set of optimum, ideal dimensions for each component footprint type and for all situations and all manufacturing processes is not possible. Any one set of patterns would

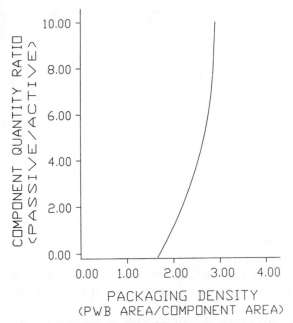

Figure 2.20 SMT packaging density determined by passive/active component ratio.

be a compromise. Wave soldering would be ideally served with one set of dimensions and reflow soldering another. The patterns shown in this book are biased toward higher reliability solder joints subjected to the more rigorous environmental exposures. See industry standards and supplier catalogs for suggested variations of footprint sizes.

CTE of solder.   When compared to other metal alloys, solder has a relatively low fatigue life. Solder's CTE (approximately 25 ppm/°C) is relatively high and therefore experiences expansion mismatch between itself and the lower CTE materials to which it is attached. Active components nearly always operate at temperatures above the PWB temperatures, and therefore the solder joints are nearly always mechanically working during operational variations.

Solder joint failures.   Failure in a solder joint, due to the stress and strain of cyclical, CTE mismatch of components, PWBs, and solder itself, starts to occur as a lateral fatigue crack beneath the component and with additional thermal cycling progresses to the outer surfaces of the joint.

Very often, complete cracks do not result in immediate electrical failures. However, they are still treated as failures. Resistance measurements are used as the arbiter to determine joint failure. Failure is defined as a tenfold increase of normalized resistance across the solder joint.

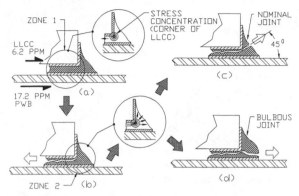

Figure 2.21   LLCC solder joint crack propagation.

Failures in LLCC joints.   There are two primary fatigue zones within LLCC solder joints. The first is the solder beneath the component, and the second is the fillet.

Solder joint fatigue cracks in LLCCs begin at the lower corner of the component castellation terminal, a natural high-stress riser point deep within the solder joint, and progresses beneath the component, in the area of lowest solder volume and highest CTE mismatch stress forces, until breaching the outer surface of the solder, thereby relieving the joint strain in that zone (see Fig. 2.21a and b).

The remainder of the solder joint, the fillet, assumes the full mechanical roll of sustaining the joint integrity. Fatigue microstructure crazing begins to spread out from the same high-stress, lower corner of the component, in all directions within the fillet. Crack progression, at this point in time, will propagate along the shortest path to final fatigue failure (see Fig. 2.21c and d).

The vertical position of lateral crack beneath the component tends to be higher when the aft edge of the PWB solder pad extends further back than the component solder termination and lower when the PWB solder pad is less (see Fig. 2.22).

Fatigue life extension of LLCC joints.   The fatigue of LLCC solder joints is greatly extended by increasing the volume of solder within the two zones affected by fatigue. Fatigue life is significantly improved when the solder height between the bottom of the component and the top of the PWB pad is greater than 0.13 mm. For maximum fatigue life, the dimension should be between 0.18 and 0.25 mm.

Fatigue life is further extended by increasing the fillet volume to the maximum extent achievable without short-circuiting to neighboring joints or disrupting soldering operations. See Fig. 2.21c and d again, and notice the difference in crack propagation between nominal solder

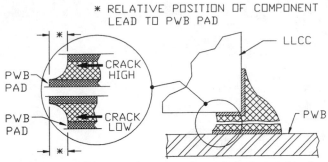

Figure 2.22   LLCC solder joint crack relative latitude—pad-lead.

joint sizes and "bulbous"-sized joints. For nominal-size fillets, the shortest path from the corner of the component to the outer surface of the solder joint is along a 45° path. For bulbous joints, the crack propagation line is parallel to the pad at the bottom of the fillet.

Pad extension for optimum-size joint.   Solder pad extension beyond the face of the component plays a critical role in the formation of the LLCC bulbous solder joint in mass reflow soldering operations. See Fig. 2.23 for the ideal pad dimension.

Passive footprints.   Passive devices are relatively simple, generally having only two leads to terminate. The footprint pattern shown in Fig. 2.24 is ideal for general placement and soldering techniques. The pattern for MELF devices is shown in Fig. 2.25.

LLCC footprints.   For LLCC and J-leaded footprint pattern, see Fig. 2.26.

Active leaded footprints.   For leaded active component footprint pattern, see Fig. 2.27.

Fine-pitched, alternate pad patterns are shown in Fig. 2.28 as a means of developing higher-yield processing.

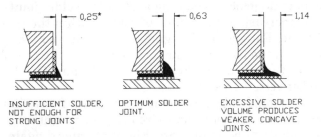

Figure 2.23   Solder pad extension optimum dimension.

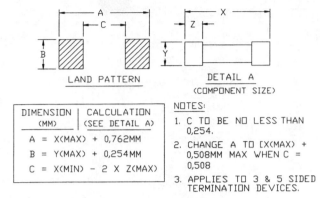

DIMENSION (MM)	CALCULATION (SEE DETAIL A)
A = X(MAX) + 0,762MM
B = Y(MAX) + 0,254MM
C = X(MIN) − 2 X Z(MAX)

NOTES:

1. C TO BE NO LESS THAN 0,254.

2. CHANGE A TO [X(MAX) + 0,508MM MAX WHEN C = 0,508

3. APPLIES TO 3 & 5 SIDED TERMINATION DEVICES.

Figure 2.24 Passive device land pattern.

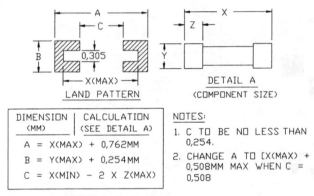

DIMENSION (MM)	CALCULATION (SEE DETAIL A)
A = X(MAX) + 0,762MM
B = Y(MAX) + 0,254MM
C = X(MIN) − 2 X Z(MAX)

NOTES:

1. C TO BE NO LESS THAN 0,254.

2. CHANGE A TO [X(MAX) + 0,508MM MAX WHEN C = 0,508

Figure 2.25 MELF device land pattern.

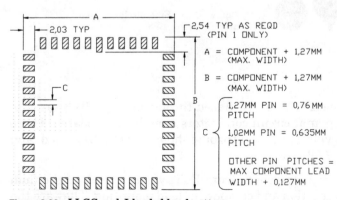

A = COMPONENT + 1,27MM (MAX. WIDTH)

B = COMPONENT + 1,27MM (MAX. WIDTH)

C {
1,27MM PIN = 0,76 MM PITCH

1,02MM PIN = 0,635MM PITCH

OTHER PIN PITCHES = MAX COMPONENT LEAD WIDTH + 0,127MM
}

Figure 2.26 LLCC and J-leaded land pattern.

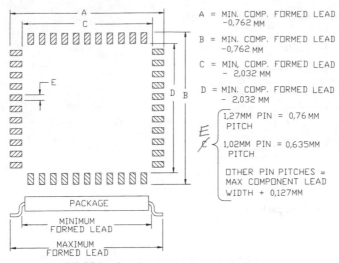

A = MIN. COMP. FORMED LEAD
    −0,762 MM

B = MIN. COMP. FORMED LEAD
    −0,762 MM

C = MIN. COMP. FORMED LEAD
    − 2,032 MM

D = MIN. COMP. FORMED LEAD
    − 2,032 MM

1,27MM PIN = 0,76 MM
PITCH

1,02MM PIN = 0,635MM
PITCH

OTHER PIN PITCHES =
MAX COMPONENT LEAD
WIDTH + 0,127MM

Figure 2.27   LDCC land pattern.

STAGGERED

INVERTED TRIANGLE

TEARDROPS

Figure 2.28 Fine-pitched   pad modifications.

2.2.7   Vias

Vias are plated holes, selectively placed throughout the PWB, used to interconnect interplanar circuit conductors in multilayer boards. There are three types of vias: (1) plated through-holes (PTH), (2) buried, and (3) blind (see Fig. 2.29).

PTH vias.   PTHs are a carryover from IMT, where they had the dual purpose of interconnection and component lead mounting. They are the only type of via used for IMT PWBs.

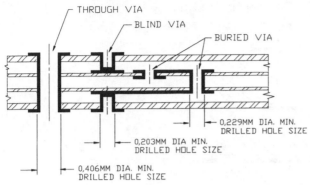

Figure 2.29   Plated hold vias.

Buried vias.   Buried vias, begun in the early 1980s for SMT interplane interconnection, are confined to internal sublaminate interconnection and do not extend to either surface of the PWB. These via types can be placed in any convenient location around the PWB; the only restriction is the avoidance of PTHs. No longer are PTHs required when interconnecting two internal circuit paths. The advantage is a far better diameter-to-thickness hole aspect ratio for manufacturing and, since buried vias and components can all be stacked on top of one another without interconnecting, packaging density is improved.

Blind vias.   These holes are relatively new and, in some circumstances, still being developed. They were introduced in the late 1980s to improve SMT density. Blind vias are external holes that do not go through the PWB. Their main function is the interconnection of external circuits, such as component footprint pads and test pads, to internal circuits. These holes can be as small as 0.15 mm in diameter and extend to the next layers beneath the top layer. As such, they can have an excellent hole aspect ratio and, more importantly, being so small, can be placed within the component pads without fear of solder runoff depleting the joint volume or thermal bleed causing insufficiently formed solder joints.

Mechanical and laser drilling.   Blind vias are drilled either mechanically or with lasers. Mechanical drilling can have depth accuracy problems, and laser drilling can involve problems with reinforcement fibers. The two approaches form two different types of holes. Deflection pads are used at the bottom of laser holes, and ordinary interplane interconnections are used for mechanically drilled holes. End mills are sometimes used for mechanical drilling. These holes look like the laser-drilled holes. This approach, however, requires close tolerancing in a wavery laminate (see Figs. 2.30 and 2.31).

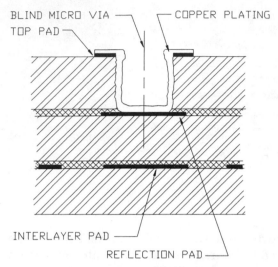

Figure 2.30 Laser-drilled blind via in SMT multilayer board.

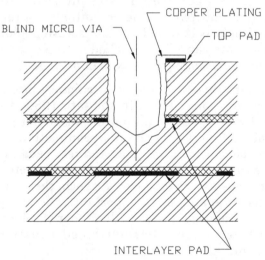

Figure 2.31 Machine-drilled blind via in SMT MLB.

Blind vias present a plating challenge, especially for holes with aspect ratios above 3:1 and diameters below 0.25 mm. Ultrasonics and other forms of plating agitation are being implemented. Blind vias in FP pads should be 0.25 mm in diameter and have an aspect ratio no larger than 2:1. Care needs to be taken, however, that the hole volume of the blind via remains significantly small and thereby does not adversely affect the solder preparations.

High-density, double-sided SMT assemblies are more readily fabricated with the use of blind and buried vias and sequential lamination of the multilayer boards.

The cost of fabricating blind and buried vias, in themselves, is relatively low. However, these via types are directly associated with sequential lamination, and it is the sequential lamination that drives up the cost of multilayer boards that contain blind and buried vias.

2.2.8 SMT reliability

Reliability of SMT is considered in two ways: (1) in comparison to IMT and (2) as an entity in itself. When comparing SMT reliability to IMT reliability, SMT must be seen on a one-for-one replacement basis. Basically, SMT components are less complex, with fewer interfaces and less materials. SMT replacement assemblies require less PWB layers, less PTHs, less interconnections, less weight, less volume, and, most importantly, less manual labor. All these features contribute to higher reliability.

Solder joints.   The area in which SMT reliability is questioned the most is the reliability of solder joints. When SMT design properly reflects the end product usage, and the manufacturing processes and materials are properly validated, the solder joints work reliably for the life-cycle duration of the product (see Figs. 2.32 and 2.33).

Solder joints in SMT are less tolerant of manufacturing process variations than are solder joints in IMT. Reliability of SMT is, therefore, far more reliant on the proper fabrication of solder joints than is IMT, es-

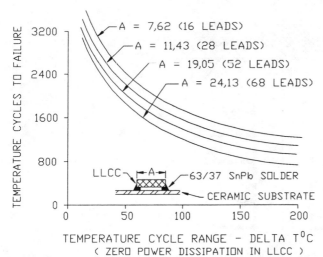

Figure 2.32 Fatigue life of LLCC solder joint as function of temperature range.

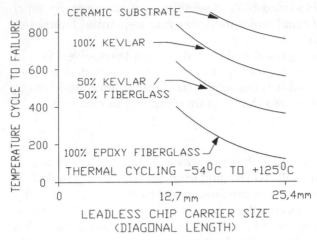

Figure 2.33 Fatigue life of LLCC solder joint for various PWB materials.

pecially for LLCC solder joints. See solder joint failure comments in Sec. 2.2.6.

Plated through-hole vias. SMT PWB materials put far more stress on PTHs than do IMT materials (see Sec. 2.2.3). The reason for these increased stresses and increased potential for PTH barrel cracking is that constraining fibers or metal sheets are placed in the *X-Y* axes of

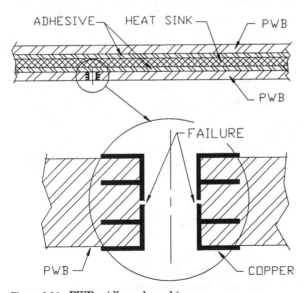

Figure 2.34 PWB midbarrel cracking.

FACTORS ON PTH STRAIN	IMPACT		
	LARGE	MODERATE	LOW
1. PWB Z-AXIS CTE	X		
2. PWB Z-AXIS ELASTIC MODULUS	X	X	
3. PWB THICKNESS	X		
4. COPPER PLATING THICKNESS	X		
5. PTH DRILLED DIA.	X	X	
6. NICKEL PLATING PTH BARREL			
a) BARREL STRAINS	X		
b) STRAIN AT PAD ROOT	X	X	
7. PTH PAD GEOMETRY			
a) LOCATION			X
b) SIZE			X
8. HEATSINK CTE			
a) WITH HARD BOND	X		
b) WITH SOFT BOND	X	X	
9. HEATSINK ELASTIC MODULUS			X
10. PWB X-Y			
a) CTE			X
b) ELASTIC MODULUS			X

Figure 2.35 Factors on PTH strain impact.

PWBs, thereby causing an inordinate amount of expansion in the Z axis (the plane of the vias) (see Fig. 2.34). Improvement in PTH reliability is gained by doubling the nominal hole wall plating thickness, by limiting the minimum hole size, and by adding a nickel overplating to the wall copper plating.

A number of factors place strain on PTHs with varying impact. Figure 2.35 shows these PTH strain factors and their influence in terms of large, medium, and low impact.

Effects on PTH strain from five different constraining fiber materials are presented in Fig. 2.36. The effects of PWB thickness on PTHs is

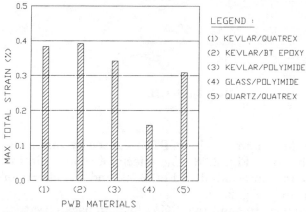

Figure 2.36 PTH maximum strain for five PWB materials.

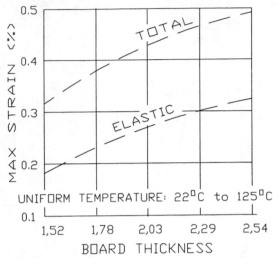

Figure 2.37   PTH maximum strain versus PWB thickness.

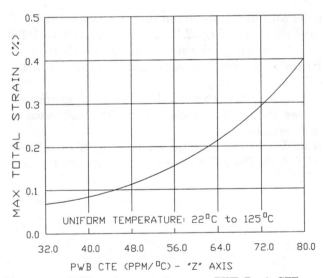

Figure 2.38   PTH maximum strain versus PWB Z-axis CTE.

presented in Fig. 2.37. Maximum strain on PTHs caused by various Z-axis CTE values is shown in Fig. 2.38. The effects of copper plating thickness on PTH strain is presented in Fig. 2.39, and the effects of PTH diameter are shown in Fig. 2.40.

The type of adhesive used to bond two CCAs back to back on either

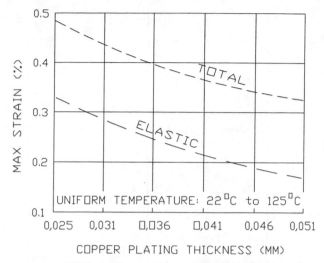

Figure 2.39 PTH maximum strain versus PTH plating thickness.

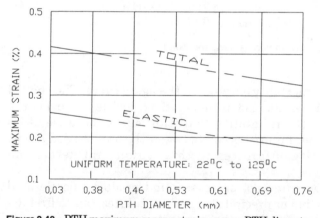

Figure 2.40 PTH maximum copper strain versus PTH diameter.

side of a constraining center plate, a plate that also serves as a heat sink, directly influences the X-Y- and Z-axis CTE of the PWB resins and consequently the PTHs. CCAs can be bonded with rigid, hard bond or soft, compliant bond. Figure 2.41 presents the effects of hard and soft adhesive bond on PTH strain.

Reliability, infant mortality, and aging.   Reliability is, in general, the measure of the useful operational life of products as defined by

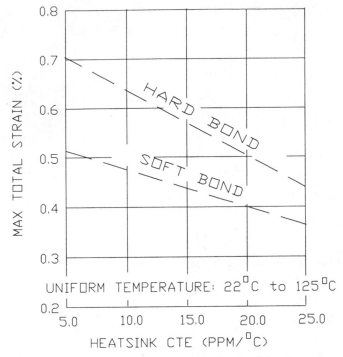

Figure 2.41   PTH maximum strain versus hard-soft bond.

combining the infant mortality curve with the aging curve. The shaded areas along both curves, as shown in Fig. 2.42, define a third curve popularly known as the reliability "bathtub" curve, because of its overall shape. Random failures occur between the two shaded areas, and it is the quantity of failures in this region that determine reliability.

In developing new materials, processes, and products, accelerated aging is used, not to find or precipitate random failures, but to first determine the wearout point. From the wearout point, the expectant life can be established and then reliability characteristics pursued.

Wearout.   Improvements can be introduced and then assessed according to how far the wearout period can be pushed to the right. Wearout in SMT mostly relates to the fatigue life of the solder joints in terms of temperature extremes and the number of thermal cycles to failure. Each segment of the industry has, more or less, its own wearout norms. Figure 2.43 presents temperature extremes and service life for major market segments in terms of expectant thermal cycles per year and the life expectancy in quantity of years.

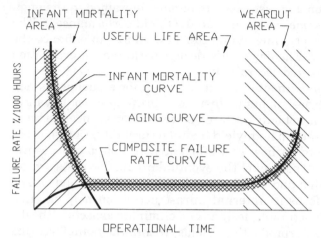

Figure 2.42 Reliability "bathtub" curve.

2.2.9 Guidelines for promoting high-yield SMT manufacturing in design

Included in the engineering's design goals and specifications for product performance should be how well each product will perform through its production cycle. It should become a matter of pride and company recognition for the design engineer whose products sail through the production line on first pass. How well products perform during their

INDUSTRY SEGMENT	Tmin C	Tmax C	ΔT C	t_D hrs	cycles/ year	years of service
CONSUMER	0	+60	35	12	365	1-3
COMPUTERS	+15	+60	20	2	1460	5 APPROX.
TELECOMM	-40	+85	35	12	365	7-20
COMMERCIAL AIRCRAFT	-55	+95	20	2	3000	10 APPROX.
INDUSTRIAL AND AUTOMOTIVE PASS. COMP.	-55	+65	20 &40 &60 &80	12 12 12 12	185 100 60 20	10 APPROX.
MILITARY GRN/SHIP	-55	+95	40&60	12	100	5 APPROX.
SPACE $\frac{leo}{geo}$	-40	+85	35	$\frac{1}{12}$	$\frac{8760}{365}$	5-20
MILITARY AVIONICS $\frac{a}{b}$	-55	+95	40-80 &20	$\frac{2}{1}$	$\frac{500}{1000}$	5 APPROX.
AUTOMOTIVE UNDER HOOD	-55	+125	60 &100 &140	1 1 2	1000 300 40	5 APPROX.

Figure 2.43 Thermal-cyclic environments per industry segments.

production fabrication and test needs to become ingrained in the engineer's psyche to the same extent that product field performance is. The engineer should do everything, no matter how subtle, to enhance the design's producibility. The engineer's design performance evaluation grade should be measured accordingly.

Engineering design without due consideration for economics is neither engineering nor design, but is, instead, a brute-force contrivance.

Design economics is not only centered on the materials used to build a product but also rooted in high-yield manufacturability. Design is the dominant determinate on how cost-efficiently an object can be produced because 70 to 80 percent of the eventual production costs are determined by the design and set in place during the design phase. The corporation should focus its internal infrastructure on helping engineering create manufacturable designs and eliminate aspects of the design that may be counterproductive. Yields, within the manufacturing activity, are among the major contributors to product costs. Manufacturing yields are determined to a great extent by intradivisional, product-related infrastructures and engineering design. Following is a list of significant guidelines that promote high-yield SMT manufacturing:

1. *Standardization.*   Standardize design features, components, materials, and manufacturing processes and procedures wherever and whenever possible. Familiarity breeds high yields.

2. *Design features.*   There are design features, such as proper component placement, that promote high-yield manufacturing, and there are other features, serving the same performance function, that hinder yields. It is incumbent on engineering, and all others who participate in the design decision loop, to learn the differences and see that the high-yield features are used whenever possible in all designs and that the low-yield features are eliminated. Line width and line spacing significantly influence PWB yields and subsequent assembly yields. Artwork plotting tolerance, photoimaging tolerances, etching tolerances, and drilling tolerances all accumulate and determine yields. Tolerances and feature sizes are cost-traded to gain the optimum balance between them for a given company. Figures 2.44 and 2.45 show two differing sets of feature sizes that have been acceptable within the industry. Design features that should be incorporated are as follows.

   a. *Component placement.*   Relative location of components has a profound interactive effect during fabrication, test, and rework. Minimum spacing between components, as shown in Fig. 2.46 for passive devices, Fig. 2.47 for leadless and J-leaded devices, and Fig. 2.48 for leaded devices must be maintained. Component orientation for wave-soldering assemblies must also be incorporated

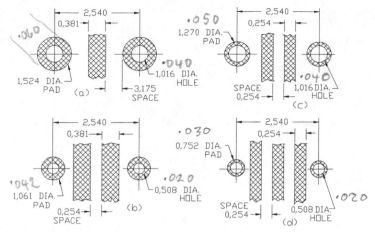

Figure 2.44 Low- and high-density line and space.

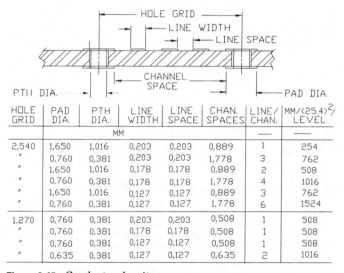

HOLE GRID	PAD DIA.	PTH DIA.	LINE WIDTH	LINE SPACE	CHAN. SPACES	LINE/ CHAN.	MM/(25.4)²/ LEVEL
		MM				—	—
2,540	1,650	1,016	0,203	0,203	0,889	1	254
"	0,760	0,381	0,203	0,203	1,778	3	762
"	1,650	1,016	0,178	0,178	0,889	2	508
"	0,760	0,381	0,178	0,178	1,778	4	1016
"	1,650	1,016	0,127	0,127	0,889	3	762
"	0,760	0,381	0,127	0,127	1,778	6	1524
1,270	0,760	0,381	0,203	0,203	0,508	1	508
"	0,760	0,381	0,178	0,178	0,508	1	508
"	0,760	0,381	0,127	0,127	0,508	1	508
"	0,635	0,381	0,127	0,127	0,635	2	1016

Figure 2.45 Conductor density.

(see Fig. 2.49). Restricted areas around footprints must also be observed (see Fig. 2.50). Components with high thermal mass should be uniformly distributed when soldered by IR processes. Optional via locations in passive land patterns are as shown in Fig. 2.51.

b. *Functional test.* Test pads should be clustered or positioned to permit ease of mass probing without misregistration or undue alignment. Test pad minimum sizes and locations, when distributed around component footprints, should be done in accordance with Fig. 2.52.

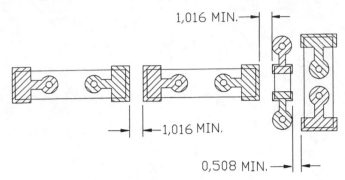

Figure 2.46 Passive land pattern spacing.

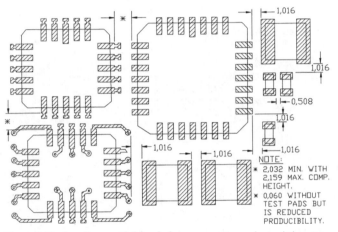

Figure 2.47 Leadless and J-leaded component spacing minimums.

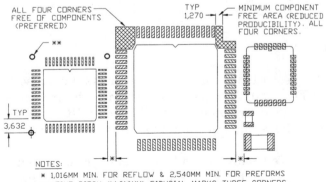

Figure 2.48 Leaded component spacing minimums.

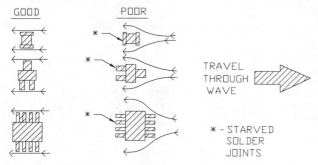

Figure 2.49 Wave-soldering component orientation.

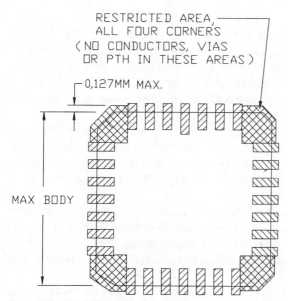

RESTRICTED AREA,
ALL FOUR CORNERS
(NO CONDUCTORS, VIAS
OR PTH IN THESE AREAS)

0,127MM MAX.

MAX BODY

Figure 2.50 LLCC restricted areas.

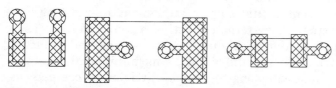

Figure 2.51 PTH locations in passive land patterns.

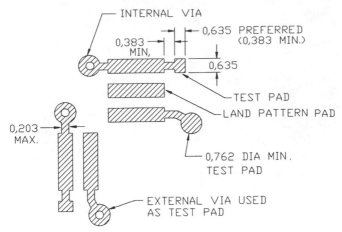

Figure 2.52   Land pattern feature limits.

All variable component adjustments on the assembly line are time-consuming touch labor and should be eliminated whenever and wherever possible. Final testing should result in a simple accept/reject determination. Although SPC data needs to be gathered during acceptance testing, interrogation of rejected circuits should be performed off-line.

c. *Component types.*   The variety of component types and component values should be kept to a minimum, especially eliminating those that add additional manufacturing steps. The ideal components are those that support a single pass of the assembly through manufacturing.

d. *Design approval.*   A SMT steering group, consisting of informed individuals representing all affected divisions within the company and who have participated in the design-related decisions beginning with the concept stage, should make a final review and approve the finished design.

e. *Failure analysis.*   In addition to the adjustments made in response to normal SPC feedback, a conscious follow-through review should be made of failed parts and failed processes to determine the root cause of the failure and gain understanding of the failure mechanism. Understanding the root cause and its mechanism is the essential first step in implementation of long-term corrective action. Long-term corrective action involves design guideline modifications and supplier feedback as well as factory process and procedure modifications. This failure analysis step is essential for improving long-term yields and for any company aspiring to reach a six-sigma (6σ) quality level. Engineering skills are often needed for this activity.

2.2.10   Material selection

Well-established materials that have served IMT and hybrids are also serving SMT (i.e., epoxy, polyamide, ceramic, aluminum, and copper). SMT, however, has been slowed by various property limitations in these established materials. Material improvements have been made and new materials developed to overcome these limitations and extend the performance of SMT. These property limitations included glass transition temperature T_g, CTE of composites, dielectric constant, weight, thermal conductivity, modulus, and water absorption. New reinforcement fibers, new resins, new alloys, and new composites have been developed and introduced that improve one or another of the limitations (see Fig. 2.53).

Fiberglass has been replaced with newer, stronger, lower CTE fibers that serve to constrain laminate CTE expansion. Aramides and quartz fibers are the two most prominent new fibers and "S" glass, the most prominent upgraded fiber.

Upgraded epoxies, multifunctional and tetrafunctional, have raised the standard material properties of standard epoxies (see examples in Table 2.1).

In tetrafunctional epoxy, an upgraded version of FR4, the T_g is raised above the standard epoxy from 125 to 140°C at a cost increase of less

SUBSTRATE MATERIALS	GLASS TRANSITION (TEMPERATURE DEG.C)	CTE (X-Y AXIS) (ppm/DEG.C)	CTE (Z AXIS) (ppm/DEG.C)	THERMAL CONDUCTIVITY (W/M DEG.C @25 DEG.C)	DIELECTRIC CONSTANT (@ 1 MHz & 25 DEG.C)	MOISTURE ABSORPTION (% of WEIGHT)
GLASS/EPOXY	125	15	48	0.16	4.0	0,10
GLASS/POLYIMIDE	260	15	58	0.35	3.5	0.32
GLASS/TEFLON	75	55	—	—	2.2	0.00
ARAMIDE/EPOXY	125	6.5	50	0.16	4.1	0.10
ARAMIDE/POLYIMIDE	225	5.0	60	0.35	3.6	1.80
ALUMINA CERAMIC	N/A	6.5	6.5	2.1	8.0	0.00
BERYLLIA CERAMIC	N/A	8.4	8.4	14.1	6.9	0.00
QUARTZ/EPOXY	125	6.5	48	0.16	3.4	0.10
QUARTZ/POLYIMIDE	270	9.0	50	0.35	3.4	0.40
POLYIMIDE/CIC CORE	270	6.5	—	57	—	0.35
EPOXY/GRAPHITE	125	7.0	48	0.16	—	0,10
POLYIMIDE/GRAPHITE	250	6.5	50	1.50	6.0	0.35
ALUMINUM-NITRIDE	N/A	6.0	—	—	9.3	—
SILICON-NITRIDE	N/A	2.8	—	—	4.0	—

Figure 2.53   PWB material properties.

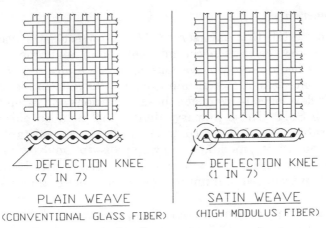

DEFLECTION KNEE
(7 IN 7)

PLAIN WEAVE
(CONVENTIONAL GLASS FIBER)

DEFLECTION KNEE
(1 IN 7)

SATIN WEAVE
(HIGH MODULUS FIBER)

Figure 2.54 Reinforcement fiber weave.

than 10 percent. Multifunctional epoxies have raised the T_g to 190°C at cost increases below 30 percent. Cyanate esters, substitutes for higher-performance epoxy applications, have a 250-T_g value along with a lower dielectric constant (3.5). Polyimides have been improved with higher peel strength and a slightly higher T_g.

Standard FR5 epoxies have been upgraded with improved moisture and chemical resistance (over two times the values in FR4), higher T_g (>200°C), and lower dielectric constant (<4.0). Traditional reinforcement cloth plain-fiber weave patterns are being studied to reduce the number of weave "knees" in each fiber strand as it runs through the weave, locking the strands together in cloth (see Fig. 2.54). By deflecting the fiber when under load of the CTE force, knees reduce the amount of CTE constraint that could otherwise be achieved with taut strands. Eliminating all knees has another major advantage; the surface of the laminate layers are much smoother, making it possible to more easily produce PWBs with higher-density conductor lines (<0.10 mm wide). Without the deflection caused by the knees in the weave, smaller-diameter filaments can be used in the weave to help reduce laminate thicknesses, further improve smoothness, improve drilling results, and reduce drilled hole sizes.

Selecting materials for SMT PWBs is unlike selecting materials for IMT boards, where a few general-purpose materials adequately serve all IMT PWB design needs. SMT PWBs have become complex and diverse. As packaging densities and circuit speeds increased, SMT PWBs have gone beyond being simply rigid component mounting planes with captured wiring, to being critical elements within the circuits themselves, where, among other things, impedance, crosstalk, and rise and fall times are helped or hindered by the materials and board configuration.

TABLE 2.2 Material Selection Criteria

Feature	Ranking
CTE (X-Y axes)	10
Technical risk	10
Resin, T_g	9
Dimensional stability	8
CTE (Z axis)	7
Availability	6
Dielectric constant	5
Water absorption	5
Machinability	5
Cost	4
Copper adhesion	4
Microcracking	4
Repairability	4
Thicknesses	3
Weight	2
Dissipation factor	2
Foreign source	2

Table 2.2 presents a group of features associated with material selection that can serve as a generic list of criteria to supplement the functional criteria in the selection process.

For high-speed circuits and chip-mounted technology (CMT—formerly known as COB for chip-on-board) SMT, PWBs will require materials with the following characteristics:

1. Dielectric constant below 3.0

2. Conductor widths below 0.127 mm

3. Conductor thicknesses below 0.013 mm

4. Dielectric thicknesses below 0.102 mm

5. Coplanar surfaces below 0.051-mm variation

6. Two-dimensional stability improvement

7. CTE below 10 ppm/°C

8. T_g above 220°C

Today's SMT PWBs are composed of a blend of old materials, modified materials, and new materials. See Fig. 2.55 for the pros and cons on some of these materials.

Two material values are unique and need further explanation beyond the reported numbers in charts: glass transition and CTE of composite structures. Copper-invar-copper is a unique material that also needs further explanation.

	CERAMIC	EPOXYGLASS	POLYGLASS	POLYKEVLAR	METAL COMP.	METAL CORE
PROS	• HI DENSITY INTERCONNECT • COMPAT. CTE • MED VOLUME PRODUCIBILITY • THERMAL CONDUCTIVITY • MED COST	• LARGE PWB • LOW COST • HI VOLUME PRODUCIBILITY • FAIR RE-WORKABILITY • LOW WEIGHT	• LARGE PWB • MED COST • HI VOLUME PRODUCIBILITY • GOOD RE-WORKABILITY • LOW WEIGHT	• LARGE PWB • MED VOLUME PRODUCIBILITY • FAIR/GOOD REWORKABILITY • LOW WEIGHT • CTE MATCH-X-Y AXIS	• LARGE PWB • MED VOLUME PRODUCIBILITY • TAILORED CTE • FAIR/GOOD REWORKABILITY • THERMAL CAPACITY • STIFFNESS	• LARGE PWB • HI VOLUME PRODUCIBILITY • TAILURED CTE • MED COST • THERMAL CAPACITY • STIFFNESS
CONS	• PRECIOUS METALS • SIZE LIMIT • POOR RE-WORKABILITY • WEIGHT • FRAGILITY • DIELECTRIC CONSTANT • DIFFICULT CKT MODS	• HI X,Y,Z AXIS CTE • THERMAL CONDUCTIVITY	• HI X,Y,Z AXIS CTE • THERMAL CONDUCTIVITY	• HI COST • MED Z AXIS CTE • ADH. FAILURE FIBER/RESIN • WATER ABSORPTION • DIFFICULT MFT'ABILITY • THERMAL CONDUCTIVITY	• WEIGHT • COST	• WEIGHT • COST OF PWB

Figure 2.55 PWB material pros and cons.

Glass transition. CTE values are predictable, linear, and approximately flat over a defined, limited temperature range. At the upper end of the linear range the CTE value radically increases. As temperatures continue to rise past that upper range point, the radical increase of the CTE can cause catastrophic damage to the PWB and assembly. This specific upper temperature point is where the mechanical and thermal properties of the material experience a drastic change, and where the molecular bonds weaken, causing the resins to go from rigid to soft. This critical point is defined as the *glass transition* and is denoted as T_g (see Fig. 2.56). Specified T_g values, however, do not represent a sudden, specific, and dramatic demarcation temperature

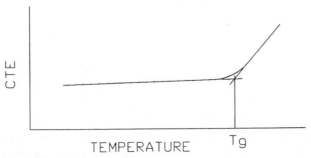

Figure 2.56 Glass transition curve.

from one material state to the next, but represent, instead, the midpoint of a very narrow interval, generally no wider than 25°C, within which the material goes through the transformation from one state to the other. A general rule, for conservative designs, is to consider the temperature point at which the initial T_g changes in material properties begin to occur to be 10 to 15°C lower than the specified T_g temperature.

CTE of composite structure. When two or more organic and/or inorganic materials are fixed to one another in intimate contact, through bonding or lamination, the resulting CTE of the composite is somewhere between the highest and the lowest CTEs of the individual materials and skewed toward the materials having the following aspects:

1. Highest tensile strength

2. Highest elastic modules

3. Lowest rosin-to-weight ratio

4. Lowest copper-to-weight ratio

5. Reinforcing fiber with the least amount of fabric weave deflections

6. Use in conjunction with constraining foil inner layers

Aramid fiber has an axial CTE of -2 ppm/°C when measured in a straight line with zero deflections and increases to 2 ppm when grouped in bundles and woven in plain-weave fabric. The axial CTE increases to 8 ppm when aramid fiber is combined with epoxy resin (CTE 60 ppm/°C) as a laminate layer and again increases to 10 ppm as a laminated PWB.

CTE measurements. Reported CTE values vary from one report to another for the same material. Variations occur between laboratories, between measurement methods, and between batches of materials.

Any one of the following five techniques is used to measure CTE:

1. Differential scanning calorimetry (DSC)

2. Thermomechanical analysis (TMA)

3. Dynamic mechanical analysis (DMA)

4. Quartz-tube dilatometry (QTD)

5. Strain gauges

No single reported CTE measurement should be considered as being authentic. Each measurement should be taken on its own merits and used judiciously.

Copper-invar-copper.   Copper-invar-copper (CIC) is a three-layer composite of two sheets of copper bonded in equal thicknesses to both sides of an invar (iron-nickel alloy) sheet using a proprietary lamination process not unlike that used in making coins. This is an ideal material for SMT design that can serve four functions simultaneously:

1. The CIC composite serves as the constraining element controlling the CTE of PWBs when bonded as either two laminated foil sheets, approximately 0.127 mm thick, within the board, or as a single, thicker sheet, approximately 1.270 mm thick, between two single-sided, Type I, SMT assemblies.

2. Laminated foils, in addition to constraining the CTE, normally serve as the ground and power planes.

3. These laminated foils help dissipate heat within the PWB. External sheets between two assemblies, in addition to serving as the CTE constraint for both assemblies, also serve as the heat sink for both.

4. These internal foils and external sheets stiffen the assemblies, often eliminating the need for a separate structural frame.

The CTE of copper-invar-copper can be tailored to match the CTE of ceramic component by altering the proportion of copper to invar. CIC with 40% copper and 60% invar results in a composite CTE of 5.8 ppm/°C.

One earlier problem with CIC was its tendency to redevelop a memory while in service, causing the PWBs to bow. The proprietary bonding process used by the sole-source supplier, Texas Instruments, uses a large drum to coin the metals, and when improperly annealed, the composite has a tendency to return to the diameter of the drum. This problem has been solved. If, however, a special order is placed for a thickness sheet or a CTE not normally produced, then the purchase order should specify consideration for annealing.

Copper-molybdenum-copper.   Copper-molybdenum-copper (CMC) is a replacement composite for CIC, serving the same functional roles. Certain material property differences, however, need to be considered when choosing one or the other.

CIC versus CMC.   CIC is the most widely used of the two materials for general applications, whereas CMC is used for high-performance applications. CMC has better thermal properties, especially in the transverse axis, and has slightly better mechanical properties. CMC is, however, heavier and costs more.

Adhesives. The choice of adhesive material used for component bonding or as standoff bumps below components must be compatible with the PWB material and not be harmed by the reflow environment. At the same time it must not induce CTE stresses in the solder joint during thermal cycling. It should also be possible to dispense adhesive materials in dots without having stringer threads trailing from the dispense nozzles and to cure the adhesive materials quickly by heat or UV in less than 5 s.

Thermally conductive films. Glass-reinforced modified epoxy adhesive films are available primarily for thermal management purposes beneath leaded components but can also serve as a component-to-PWB standoff device. This film is available in thicknesses ranging from 0.051 to 0.254 mm and can be precut into appropriate sizes.

Material validation. Components, PWBs, and expendables (solder paste, chemicals, etc.) should all be validated for compliance to specifications prior to placement on the assembly line. Values, burn-in tests, solderability, characteristic reactions, and other factors need to be verified while in the inventory cycle.

New materials and components should go through a one-time generic qualification cycle and from then on, as incoming materiél, be validated to verify compliancy with the original, qualified item.

2.2.11 Impact of PWB design on electrical performance

Since SMT packaging densities make it far more practical for high-performance circuits to be incorporated on standard SMT PWBs than on IMT PWBs, these circuits are more likely to be encountered in SMT system designs. Performance driven designs use high-speed circuits that can have signal pulse rise times in the submicro region and some state-of-the-art circuit rise times in the subnanosecond region. Preserving the signal edge waveform requires low path resistance, low inductance, and short signal path lengths. At high clock rates, signal reflections and ringing can affect waveforms. Signal pulse time roundtrip return needs to be kept within 25 percent of the signal rise time if the high-performance circuits are to interactively operate properly.

Parasitic line capacitance and conductor geometries can inhibit signal quality and circuit timing contributing to propagation delays, vertical and lateral crosstalk, and characteristic impedance, all of which are interactive.

Routing high-performance signal lines cannot be treated as normal, general-purpose PWB wiring. Dielectric constant of the laminated insulation material, insulation thicknesses, conductor cross-sectional area, signal-to-signal and signal-to-ground orientations, and conductor lengths all must be controlled by the PWB design.

In addition to careful selection of PWB materials to control the electrical characteristics of PWBs, high-speed signal paths must be treated as transmission lines (see Fig. 2.57).

Characteristic impedance.   Characteristic impedance control in fixed geometry PWB designs is a matter of maintaining a uniform dielectric constant, accounting for such things as parasitic capacitive loads of via holes (1.0 pF) and crossing signal lines (0.01 pF) and maintaining signed path positions relative to the reference planes.

Propagation delay.   Timing delays between circuit gates is due, in part, to the length of the signal paths. Overall delay results from the combined length of on-board conductors and length of signal path within components. SMT has made it possible to significantly reduce circuit path lengths within components (see Fig. 2.58). Until recently, delay within the component was the major contributor to overall signal delay. Although SMT packaging densities have caused reduction of the on-board circuit paths, PWBs now are the major contributor to overall delay (see Fig. 2.59).

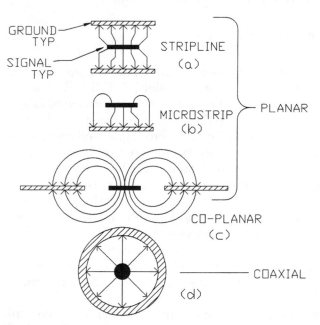

Figure 2.57   Transmission line types.

COMPONENT LEAD COUNT	RATIO				
	LONGEST TO SHORTED TRACE		RESIST.	INDUCT.	CAP.
	IMT	SMT	IMT TO SMT		
16-18	6 : 1	1.5 : 1	2 : 1	6 : 1	6 : 1
24-32	3 : 1	1.5 : 1	4 : 1	8 : 1	8 : 1
40-44	6 : 1	1.5 : 1	5 : 1	9 : 1	8 : 1
64-68	7 : 1	1.5 : 1	6 : 1	8 : 1	8 : 1

Figure 2.58   Ratio of electrical characteristics—IMT versus SMT.

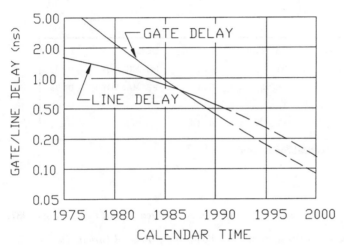

Figure 2.59   Gate-line delay gap.

Various materials also play a role in delaying the signals due to dielectric constant and electrical conductivity characteristics (see Figs. 2.60 and 2.61).

Crosstalk.   Crosstalk, electrical coupling between conductor lines due to capacitive and electromagnetic mechanisms, can be controlled by separating the lines, adding shielding, or placing ground lines between conductors and ground planes above and/or below, thus reducing parallel lengths to no more than half the distance a signal pulse would travel in one rise-time period.

SUBSTRATE MATERIAL	DIELECTRIC CONSTANT (Er)	PROPOGATION DELAY (NANOSEC./0.31M)
VACUUM (BASE DATA)	1.0	1.0
TEFLON-GLASS	2.2	1.5
EPOXY-KEVLAR	3.6	1.9
POLYIMIDE-QUARTZ	4.0	2.0
EPOXY-GLASS	4.7	2.2
ALUMINA (96%)	10.0	3.2
THICK-FILM DIELECTRICS	8-12	2.8-3.5

Figure 2.60   Typical propagation delays for substrate material.

SILVER	1.05	CADMIUM	0.23	LEAD	0.08
COPPER	1.00	NICKEL	0.20	MU-METAL	0.03
GOLD	0.70	IRON	0.17	PERMALLOY	0.03
ALUMINUM	0.61	TIN	0.15	STAINLESS	0.02
MAGNESIUM	0.38	STEEL	0.10		

Figure 2.61   Relative electrical conductivity.

Bibliography

Anderson, James: "A Look at Surface Mount Basics," *Printed Circuit Design*, July 1987, pp. 39–40.

Blankenhorn, James C.: "In Search of SMT Footprints," *Printed Circuit Design*, May 1992, pp. 10–15.

Capills, Carmen: *Surface Mount Technology*, McGraw-Hill, New York, 1990.

Coombs, Clyde F., Jr.: *Printed Circuits Handbook*, McGraw-Hill, New York, 1988.

Cox, Mark R., and Lee H. Ng: "An Economic Comparison of High and Low Temperature Cofired Ceramics," *Inside ISHN*, January 1992, pp. 31–36.

Derman, Glenda: "Better Materials Match Performance of Boards," *Electronic Engineering Times*, February 24, 1992, pp. 39–49.

Electronic Packaging Handbook: The Institute for Interconnection and Packaging Electronic Circuits (IPC).

Engelmaier, Werner: "Functional Cycling and Surface Mounting Attachment Reliability," *ISHM Technical Monograph Series*, October 1984, pp. 87–114.

Grzesik, Tony: "Top Down Design," *Printed Circuit Design*, June 1992, pp. 10–14.

Harper, Charles A.: *Electronic Packaging and Interconnection Handbook*, McGraw-Hill, New York, 1991.

Hollomon, James K., Jr.: *Surface Mount Technology for PC Board Design*, Howard W. Sams & Co., Indianapolis, Ind., 1989.

Korf, Dana: "High Density SMT Using Laser Blind Vias," *Surface Mount Technology*, June 1988, pp. 35–36.

———, "Laser-Drilled Blank Vias Increase PCB Real Estate," *Electronic Packaging and Production*, February 1987.

Kotlowitz, R. W., and Werner Engelmaier: "Impact of Lead Compliance on the Solder Attachment Reliability of Leaded Surface Mounted Devices," *Proceedings of*

International Electronics Packaging Society Conference, San Diego, Calif., November 17–19, 1986, pp. 841–865.

Lampe, John W.: "The Interrelationships of Design, Materials, and Processes for Surface Mount Assemblies in Military Applications," *Surface Mount Technology,* November 1989, pp. 22–27.

Lynch, Ferrence P.: "Get a Handle on Heat," *Design News,* January 21, 1991, pp. 102–104.

Meusey, Jim: "Plastic SM Packages Improve," *Electronic Engineering Times,* September 25, 1989, pp. 56, 62, 68.

Mollison, Mary: "Blind and Buried Via Fabrication," *Electronic Packaging Production,* January 1987, pp. 106–107.

Murray, Jerry: "Blind Buried Vias," *Circuits Manufacturing,* April 1988, pp. 62–64.

Murray, Jerry: "PCB Laminate Overview," *PC FAB,* February 1992, pp. 44–50.

Pattison, William: "In-plane CTE Determinations of High Performance Materials within the Manufacturing Technology for Advanced Data Signal Processing Program," IPC-TP-879, paper presented at IPC 33d Annual Meeting, April 2–6, 1990.

Prasad, Ray P.: "Designing Surface Mount for Manufacturability," *Printed Circuit Design,* May 1987, pp. 8–11.

Pukaite, Clifford J., and Anton B. Usouski: *Surface Mount Technology,* June 1988, pp. 39–40.

Robinson, Gail M.: "Heat Transfer Strip Keeps Electronics Cool," *Design News,* February 10, 1992, pp. 196–197.

Snadeau, Rene F.: "Solution to the SMT Maze," *International Electronics Packaging Society Inc. Journal,* vol. 9, no. 3, 1988, pp. 14–16.

Solberg, Vern: "SMT Land Pattern Standards," *Printed Circuit Design,* September 1990, pp. 23–31.

Swagerm, Anne Watson: "Circuit Design Requires Thermal Expertise," *EDN,* June 22, 1989, pp. 93–104.

Veltzen, Ken: "Pushing the Packaging Envelope," *Circuit Assembly,* March 1992, pp. 30–35.

Walcutt, Jim: "Managing the Thermal Tradeoffs," *Circuit Assembly,* March 1992, pp. 36–41.

Winkler, Ernel R., and Karl W. Rosengarth, Jr.: "Take a System Approach to High Density VLSI Packaging," *Integrated Circuits Magazine,* March 1985, pp. 55–60.

Woodgate, Ralph W.: "Weak Knees and Design," *PC FAB,* January 1991, pp. 56–60.

3

Components

Surface mount technology derives its name from the way its components are attached to printed wiring boards (PWBs). In SMT active and passive circuit elements are packaged in a variety of shapes, sizes, and configurations with one common feature among them; each is mechanically attached and electrically connected to the PWB through their multiple leads soldered to matching, coplanar pads on one or both surfaces of the PWB. Single-plane attachment makes it possible, and convenient, to subminiaturize components and to automate the fabrication of the SMT assemblies.

3.1 Three Component Categories

Traditionally, there are three categories of SMT components: (1) active, (2) passive, and (3) others. Active components are semiconductor devices that amplify, switch, or rectify electronic signals. Included in this category are transistors, diodes, and integrated circuits (ICs). Passive components are conductive devices that alter signals without amplifying, switching, or rectifying. Resistors, capacitors, and inductors are prominent examples of components in this category. The third category consists of conductive devices that simply pass signals, unaltered by the device itself, along prescribed pathways from one functional circuit to another.

The traditional term "passive," for the middle category, is a misnomer. In retrospect, it would have been somewhat more accurate to have labeled the three categories as (1) active, (2) semiactive, and (3) passive. The last category is the category that passively passes the signal along unaltered. This volume, in keeping with tradition, however, will reluctantly continue to use the term "passive" as the descriptor for the middle class of components and "others" as the descriptor for the third category.

In most components, the attachment of SMT devices is made solely with, and totally dependent on, the solder attachment to the terminals on the devices themselves. Since so much depends on component terminals and the methods of attaching them, selection of terminal configuration and finish dominates design and assembly procedural development. Terminals are, therefore, an essential issue and a good starting point for any discussion on SMT components.

3.2   Component Terminals

Electrical and mechanical attachment is the primary role served by SMT component terminals. A secondary role for leaded terminals, and almost as important as the first role, is the leaded terminal's ability to furnish the necessary compliance between the solder joint and the PWB and to help protect the solder from fatigue failure caused by CTE mismatch within the joint. How well terminals perform all these duties determines, to a great extent, the reliability of solder joints and, therefore, that of the end product itself.

Selecting the proper terminal type for the product application is a prerequisite to achieving the most cost-efficient and reliable SMT design. The ideal terminal is sufficiently rugged to withstand all routine handling and yet compliant enough to easily compensate for the worst case, CTE mismatch of materials. Compliance of the ideal terminal, when lined up in rows along two or four sides of the component, should be uniformly in the transverse as well as the X and Y directions.

Two general groups of terminals are used on surface mounted components: (1) leaded and (2) leadless. Component terminations that are formed independent of, and projected away from, the component body are grouped as leaded. Component terminations formed by metallizing portions of the component body surfaces are grouped as leadless. Passive components, with very few exceptions, have leadless terminations, whereas active components can have either leadless or leaded terminations.

3.2.1   Passive component lead style

Pin, lead, and *terminal* are terms often used interchangeably in referring to both leadless and leaded terminations on SMT components.

Leadless terminations on passive devices, often referred to as "end caps," are formed by first applying a fused, metallization layer of silver-palladium, or another equivalent alloy. Silver-palladium is ideally suited to effect the basic electrical contact with the working element of the device. This end-cap layer is often followed by a barrier coat of

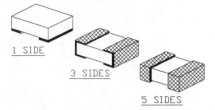

Figure 3.1 Chip termination types.

nickel to protect the inner layer of silver from being leached by the tin in the solder during the assembly soldering process. A final coat of tin-lead is often applied over the nickel to improve solderability and protect the termination from oxidizing into a nonsolderable surface. Maximum long-term solderability of the terminations can be gained by fusing the outer, tin-lead, coat. Adding the nickel and fusing the tin-lead coating, however, require additional effort by suppliers and consequently add costs to the components. Passive chip terminations are available as single-sided, three-sided, or five-sided (see Fig. 3.1).

3.2.2 Active device lead styles

Active surface mounted devices are available with different termination styles. There are several dominant styles used on leaded packages and one style on leadless devices. This leadless style of termination on active devices is different from the leadless style of termination on passive devices. On active devices, the leadless termination is formed by metallizing concaved surfaces of vertical flutes positioned along the outer edges of the component body with connecting metallized foot pads on the bottom surface.

Leaded terminations are identified by their particular shapes: gull wing, J lead, C lead, and I lead (often referred to as a *butt lead*) (see Fig. 3.2).

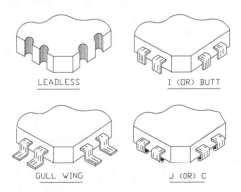

Figure 3.2 SMT lead configurations.

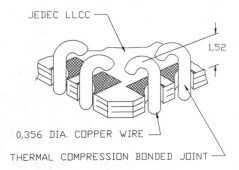

JEDEC LLCC

1,52

0,356 DIA. COPPER WIRE

THERMAL COMPRESSION BONDED JOINT

Figure 3.3 LLCC modified with leads.

Active devices can be converted from leadless to leaded by affixing adaptive leads. Adaptive leads can be affixed to leadless devices by either the user or the supplier.

It is sometimes necessary to resort to these add-on leads to compensate for the lack of availability of leaded components in the desired functional type or when upgrading an existing leadless design to one that requires a more rigorous environment component lead compliance.

There are several configurations of adaptive, add-on leads. One supplier has developed a way to thermally compression-bond round or rectangular copper leads to the castellation on the outer perimeter of leadless components (see Fig. 3.3). Leadless components are often drop-shipped by the user to this supplier for the lead adaptation for less total cost than would be incurred by starting with leaded devices. Another supplier provides tiny, spiral, copper columns filled with eutectic solder to be simultaneously attached to the bottom surface of the leadless termination. Other companies supply adaptive clip-on-leads, while others supply multicurved leads for bottom attachment (see Fig. 3.4).

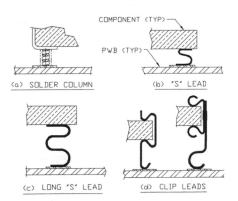

COMPONENT (TYP)

PWB (TYP)

(a) SOLDER COLUMN

(b) 'S' LEAD

(c) LONG 'S' LEAD

(d) CLIP LEADS

Figure 3.4 Add-on compliant leads.

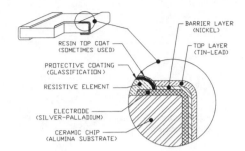

BARRIER LAYER
(NICKEL)

RESIN TOP COAT
(SOMETIMES USED)

TOP LAYER
(TIN-LEAD)

PROTECTIVE COATING
(GLASSIFICATION)

RESISTIVE ELEMENT

ELECTRODE
(SILVER-PALLADIUM)

CERAMIC CHIP
(ALUMINA SUBSTRATE)

Figure 3.5 Chip resistor construction.

3.3   Component Body Construction and Materials

3.3.1   Passive resistors

One of the least complex and most commonly used component constructions in SMT is the small, rectangular block of alumina ceramic, referred to as a *chip* and used as the primary body element for fixed resistors. Starting with these bare ceramic chips and adding a screened thick-film layer of ink material, generally made of ruthenium oxide or a diffused thin-film layer and then adding two end-cap terminals, is all that is necessary to produce chip resistors (see Fig. 3.5).

3.3.2   MELF resistors

Construction of MELF devices is similar to the construction of axial lead resistors, except there are not axial leads and the terminations are formed by metallizing each end.

3.3.3   Passive capacitors

Monolithic ceramic capacitors are the most widely used capacitors in SMT. Because of their body shape, these devices, like fixed SMT resistors, are referred to as *chip capacitors*; however, they are not constructed from small solid blocks of ceramic. These devices consist of multiple layers of thin ceramic sheets alternately printed with metal electrodes (see Fig. 3.6). The stacked layers are sintered at elevated temperatures to form the monolithic construction. Terminals are then formed on each end by a series of metalized layers.

Capacitor values and tolerances over temperature are determined by the quantity of dielectric-electrode layers and the type of ceramic. Since the ceramic sheets serve as the dielectric insulation between the layers of electrodes as well as the basic component body structure, the electrical characteristics of the particular ceramic selected have a direct effect on the electrical characteristics of the capacitor.

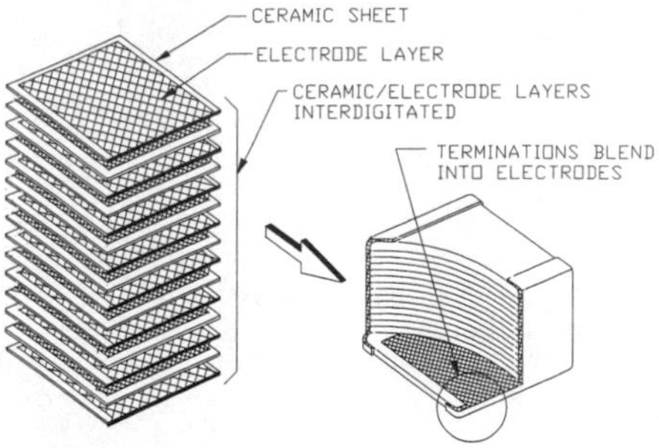

Figure 3.6 Ceramic chip capacitor construction.

   In their dielectric role, ceramics used for these capacitors fall into three general categories:
   1. *COG* (*NPO*).   This is the most stable capacitor dielectric available with linear temperature coefficients. Capacitance values vary by ±30 ppm during changes in temperature from −55 to 125°C. This type of dielectric material, however, has a relatively low dielectric constant, in the range of 80, and therefore the range of capacitance values is low, 1.0 pF to 0.027 μF, compared to that of other categories of ceramic capacitors. Capacitance change with frequency is minimal. This type of ceramic has little or no aging with time.
   2. *X7R* (*BX*).   This is the most stable capacitor dielectric available within the nonlinear temperature coefficient ranges. Capacitor values vary by plus or minus 15 percent during changes in temperature from −55 to 125°C. This type of dielectric material has a dielectric constant, in the range of 1400, and a subsequent range of capacitance values 100 pF to 1.2 mF.
   3. *Z5U*.   This category of dielectric is referred to as the "general-purpose," low-cost ceramic where temperature applications have limited impact and cost is important. Capacitors within this category of high dielectric constants provide the highest capacitance of the three but with a very steep variation (+80 to −20 percent). Capacitance value range is 2700 pF to 2.7 μF. Despite their instability, however, this category is popular because of the relatively small size and an excellent frequency response for recouping in applications where close capacitance value is not required. (*Note:* Another dielectric type within this category is Y5V, which is characteristically very close to Z5V. Capacitance values in this type vary from 2200 pF to 3.3 μF.)
   Tantalum capacitor construction begins with a pure tantalum powder pressed and sintered around a tantalum wire, resulting in a very porous structure providing an anode with a large surface area. A dielectric layer of tantalum pentoxide is anodized on the porous surfaces. The cathode is formed by layers of manganese dioxide pyrolytically processed over the dielectric layer. The catlock terminal is then formed by

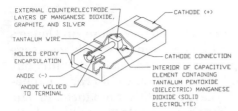

Figure 3.7 Tantalum capacitor construction.

deposition of graphite over the manganese dioxide layer followed by a silver coating and then a conductive adhesive layer (see Fig. 3.7).

3.3.4 Active

Active devices are available in two basic package types: (1) nonhermetically sealed, molded plastic packages and (2) hermetically sealed packages made of ceramic, glass, or a combination of the two. Beginning with a lead frame, the non–hermetically sealed component body can be premolded and the active element added later, or the body can be postmolded after the active element has been attached to the lead frame (see Fig. 3.8). This process is not unlike that for the dual-in-line packages of the earlier technology. Hermetically sealed packages generally consist of ceramic sheets containing selected cutouts and interconnection, and sealing metalization features are cofired into a monolithic entity (see Fig. 3.9).

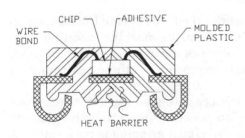

Figure 3.8 Molded plastic package.

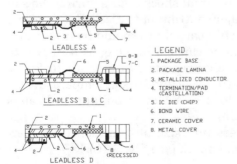

Figure 3.9 JEDEC LLCC construction.

3.4 Component Reliability

Ceramic chip capacitors and resistors have been used more, and longer, than any of the other SMT components. Their annual worldwide usage is measured in the billions. When properly applied, their reliability has met all requirements. Leadless and leaded ceramic chip carrier reliability has also been proved over the years in military products. Plastic leaded chip carriers (PLCCs), although relatively new, have been proved by industrial and MIL-STD-883 tests to be slightly more reliable than their counterpart IMT, plastic, dual-in-line packages (DIPs). Although PLCCs and plastic DIPs use the same materials and have the same chips embedded, the PLCC reliability advantage is due to its smaller size and consequential less absolute stress caused by internal shrinkage. In addition, there have been some minor improvements in the PLCC design that also help account for the better reliability.

3.4.1 PLCC cracking

Plastic components absorb moisture over time, and consequently if the moisture is not baked out prior to soldering, the component body will crack when exposed to sudden temperature rises during reflow soldering. When prebaked and preheated, these components can be safely reflow-soldered multiple times. (A maximum of five reflow cycles is recommended.)

3.4.2 Ceramic chip capacitor cracking

Unlike IMT ceramic capacitors, where the leads isolate the component from stresses caused by board flexing during machine insertion, testing, and handling and from thermal shock during wave soldering, SMT ceramic chip capacitors are directly exposed to all these environments without the benefit of leads.

In addition to the conventional forces acting on capacitors there are two other forces unique to SMT. Component gripping jaws on pick and place machines can create impact damage if gripping is performed too fast, jaw pressure is excessive, or thermal shock occurs during the soldering operation—which is the major contributor to cracked capacitors. Ideally, temperatures of ceramic chip capacitors should be raised at a rate of 3°C/s, although some ceramic capacitors have been known to consistently survive temperature changes at a rate of up to 7°C/s. Vapor-phase soldering can raise parts from 25°C (room temperature) to 217°C (vapor temperature) in 1 to 2 s. A slower buildup of heat is needed to prevent capacitor cracking.

Ceramic chip capacitor thickness and size also influence the percentage of cracks caused by thermal shock (see Fig. 3.10).

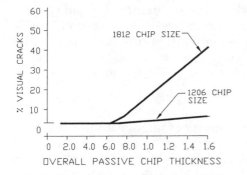

Figure 3.10 Passive chip thermal shock cracks and chip thickness.

3.5 Shipping, Storing, and Handling Packages

When purchasing SMT components, the configuration of the component shipping containers cannot be taken for granted or left to the discretion of component suppliers. To facilitate delivery and enhance the benefits of just-in-time (JIT) inventory, the original shipping package should serve as the in-plant storage and handling container. These containers, with their components, should be capable of being directly attached to production assembly machines, thereby reducing reloading efforts and changeover downtime internals.

Deterioration of component lead solderability and deformation of lead shapes are two major concerns for SMT components during shipping, storage, and handling. Mechanical, chemical, and atmospheric environmental protection should be accomplished, to a large extent, by the component shipping container itself.

3.5.1 Packaging form

Component containers designed for shipping, storage, and handling, referred to by component suppliers as *packaging,* are manufactured in several forms: tape and reels, waffle trays, feeder tubes, and bulk. Tape and reel is the preferred method for large-quantity production, because of its inherent ability to contain larger quantities of parts, lower required downtime for reloading, and controlled orientation and position of parts. Bulk systems are also used for large-scale production but are suited more for low-density component assemblies normally encountered on consumer products, where orientation and precision placement of parts are not critical. Large-scale production, however, is not the norm for the majority of SMT manufacturers; rather, smaller production runs with customized products are becoming increasingly the norm. Under these conditions, where lower quantities, rapid change-over, and flexibility of production setup becomes necessary, waffle packs and feeder tube sticks are often used. Waffle packs and feeder tubes, however, are not ideal where production throughput, regardless of quantity, makes the difference between profit and loss.

3.5.2 Transfer packaging

What is currently needed by the majority of manufacturers is a technique that enhances production throughput, is easily adaptive to a rapidly changing mix of components, is a single container with a large quantity of oriented parts, and is compatible with a large variety of existing production machines.

Traditional tape-and-reel methods are not ideally suited to fulfill the current flexibility needs of SMT manufacturers for the following reasons:

1. With traditional tape-and-reel methods, single component types with single values are loaded by the thousands on single reels.

2. Many parts used on current products have near-term limited usage.

3. Components are obtained from many different suppliers.

4. Not all suppliers are capable of putting their components on tape and reel.

5. Often, due to costs, parts on tape and reel are supplied only for orders above certain minimum quantities, quantities that far exceed near-term usage.

A relatively new approach to packaging components is now available through specialty taping machines that make it possible for users to transition from the traditional tape-and-reel packaging system to a more responsive system that meets current, lower-quantity, flexibility needs. Customized single reels, with variable component values and component types to match specific products, can now be made up by either the assembly manufacturer, the contract taping service company distributors, or the component suppliers, whichever offers the most advantageous cost and schedule response.

Add-on features such as vision inspection, laser marking, and testing are also now available which can be integrated in these newer, customized, taping machines. These add-on features help manufacturers come closer to the goal of a 0-ppm defect level for all components at time of assembly. In addition to reel-to-reel retaping, there are add-on systems to retape components from any one of the other methods, such as tubes, trays, or bulk to tape (see Fig. 3.11).

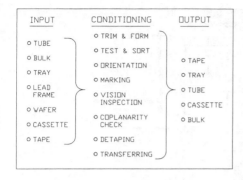

INPUT	CONDITIONING	OUTPUT
○ TUBE	○ TRIM & FORM	
○ BULK	○ TEST & SORT	○ TAPE
○ TRAY	○ ORIENTATION	○ TRAY
○ LEAD FRAME	○ MARKING	○ TUBE
○ WAFER	○ VISION INSPECTION	○ CASSETTE
○ CASSETTE	○ COPLANARITY CHECK	○ BULK
○ TAPE	○ DETAPING	
	○ TRANSFERRING	

Figure 3.11 Transfer packaging of SMT components.

3.5.3 Standard tape-and-reel configurations

Standard tape-and-reel sizes and specific features are governed by Standard Electronic Industry Association (EIA) specification number EIA481-A (see Fig. 3.12). Standard carrier tape sizes are 8, 12, 16, 24, 32, 44, and 56 mm. The two smallest size tapes can be constructed either from the less costly punched cardboard or plastic tape having the same thickness as the component or from embossed plastic, having pockets for each component (see Figs. 3.13 to 3.16). All larger-size tapes are made with embossed plastic only. A sealed or mechanically locked cover tape is used for environmental- and tamper-proofing purposes

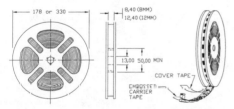

Figure 3.12 8-mm/12-mm reel per EIA Standard RS-481.

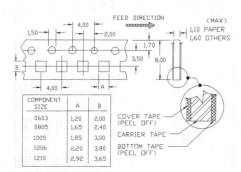

COMPONENT SIZE	A	B
0603	1,20	2,00
0805	1,65	2,40
1005	1,85	3,00
1206	2,20	3,80
1210	2,92	3,65

Figure 3.13 8-mm punched (paper) tape dimensions.

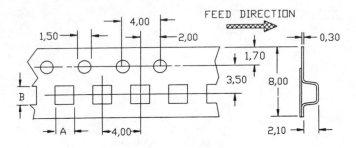

COMPONENT SIZE	A	B
0603	—	—
0805	1,55	2,35
1005	—	—
1206	2,00	3,60
1210	2,90	3,60

Figure 3.14 8-mm embossed tape dimensions.

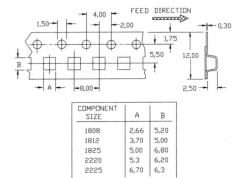

COMPONENT SIZE	A	B
1808	2,66	5,20
1812	3,70	5,00
1825	5,00	6,80
2220	5,3	6,20
2225	6,70	6,3

Figure 3.15 12-mm embossed tape dimensions.

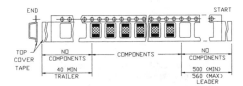

Figure 3.16 8-mm tape leader-trailer.

once the components are inserted into the embossed pockets of the tape carrier. Covers can be made clear for ease of inspection.

Components with large lead quantities and/or fine-pitched leads are still supplied in waffle trays: tape and reel for these components are not yet available.

3.6 Passive Components

Passive devices consist of *electronic devices* such as resistors, capacitors, inductors, transformers, filters, fuses, delay lines, crystals, and other derivative-type devices.

Because of their small size, most of the discrete SMT passive devices cannot be marked individually with part numbers and values. Various color codes and abbreviations have consequently been derived for these purposes by the manufacturers. Refer to the individual manufacturer's literature for a description of these codes.

3.6.1 Resistors

Chip resistors. A wide variety of SMT resistors is available from many manufacturers. The most common type used is the low-power rectangle-shaped chip resistor. This type of resistor is available in standard resistance values and tolerances (Fig. 3.17) and standard package sizes (Fig. 3.18). These resistors are manufactured in two basic types, thick-film and thin-film. Both are formed on alumina ceramic chips as shown in Fig. 3.5. Final resistance value adjustment is achieved by laser-trimming a portion of the resistive element midway between the terminal (see Fig. 3.19).

Thick-film resistors. Thick-film resistors are, by far, more common than thin-film resistors because of their relatively low cost. Typical thick-film tolerance ranges between 5 and 1 percent, making this resistor type more suitable for general-purpose applications.

VALUES FOR 2% AND 5% TOLERANCES									
1.0	1.1	1.2	1.3	1.5	1.6	1.8	2.0	2.2	2.4
2.7	3.0	3.3	3.6	3.9	4.3	4.7	5.1	5.6	6.2
6.8	7.5	8.2	9.1						
VALUES FOR 1% AND LESS TOLERANCES									
100	102	105	107	110	113	115	118	121	124
127	130	133	137	140	143	147	150	154	158
162	165	169	174	178	182	187	191	200	205
210	215	221	226	232	237	243	249	255	261
267	274	280	287	294	301	309	316	324	332
340	348	357	365	374	383	392	402	412	422
432	453	464	475	487	499	511	523	536	549
562	576	590	604	619	634	649	665	681	698
715	732	750	768	787	806	825	845	866	887
909	931	953	976						

Figure 3.17 Standard resistor decade values.

SIZE CODE	L	W	T
INCH/MM	INCH/MM	INCH/MM	INCH/MM
0402/1005	.040/1,00	.020/0,50	.020/0,50
0504/1210	.050/1,27	.040/1,02	.040/1,02
0603/1508	.060/1,52	.030/0,075	.035/0,90
0805/2012	.080/2,00	.050/1,27	.055/1,40
1005/2512	.100/2,54	.050/1,27	.059/1,49
1206/3216	.126/3,20	.060/1,52	.059/1,49
1210/3225	.126/3,20	.100/2,54	.059/1,49
1505/3812	.150/3,81	.050/1,27	.050/1,27
1805/4512	.180/4,57	.050/1,27	.059/1,49
1808/4520	.180/4,57	.080/2,00	.080/2,00
1812/4532	.180/4,57	.126/3,20	.080/2,00
1825/4564	.180/4,57	.252/6,40	.080/2,00
2018/5145	.200/5,08	.180/4,57	.025/0,64
2110/5325	.210/5,33	.098/2,49	.025/0,64
2218/5745	.220/5,72	.175/4,45	.070/1,78
2220/5650	.220/5,60	.220/5,60	.070/1,78
2225/5664	.220/5,60	.250/6,40	.080/2,00
2321/5853	.230/5,84	.210/5,33	.060/1,54

(INCH) $\frac{.012}{.028}$

(MM) $\frac{0,30}{0,70}$

TERMINAL (LEAD)

LASER TRIM ZONE

RESISTIVE ELEMENT

CERAMIC CHIP

Figure 3.18 Standard chip resistor dimensions.   **Figure 3.19** Laser trimmed resistance value.

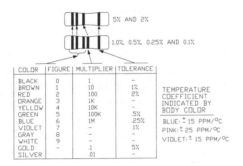

5% AND 2%

1.0%, 0.5%, 0.25% AND 0.1%

COLOR	FIGURE	MULTIPLIER	TOLERANCE
BLACK	0	1	–
BROWN	1	10	1%
RED	2	100	2%
ORANGE	3	1K	–
YELLOW	4	10K	–
GREEN	5	100K	.5%
BLUE	6	1M	.25%
VIOLET	7	–	.1%
GRAY	8	–	–
WHITE	9	–	–
GOLD	–	.1	5%
SILVER		.01	–

TEMPERATURE COEFFICIENT INDICATED BY BODY COLOR

BLUE: $\pm$ 15 PPM/°C
PINK: $\pm$ 25 PPM/°C
VIOLET: $\pm$ 15 PPM/°C

Figure 3.20 MELF resistor value and tolerance markings.

Thin-film resistors.   Thin-film resistors have a lower and more linear temperature coefficient of resistance (TCR), higher stability, and a tolerance capability suitable for precision-value applications. Thin-film tolerance ranges between 1.00 and 0.05 percent.

Metal-electrode faced (MELF) resistors.   MELF resistors are cylindrically shaped carbon-film and metal-film resistors. They are available in a wide variety of values and tolerances (see Figs. 3.20 and 3.21).

Power resistors.   Wirewound power resistors with power ratings of up to 3.5 W are available in wrapunder, J-type, leaded packages (see Fig. 3.22) and up to 5.0 W in gull-wing-type leaded packages (see Fig. 3.23).

Variable resistors.   There are two basic types of variable resistors:

1. Those completely enclosed in sealed packages that protect the resistive element against contamination due to harsh environments or molten solder during assembly procedures (see Figs. 3.24 and 3.25).

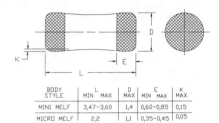

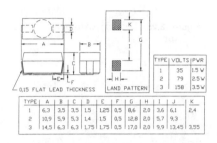

BODY STYLE	L MIN MAX	D MAX	E MIN MAX	K MAX
MINI MELF	3,47–3,60	1,4	0,60–0,85	0,15
MICRO MELF	2,2	1,1	0,35–0,45	0,05

Figure 3.21 Precision miniature MELF resistors.

TYPE	VOLTS	PWR
1	35	1.5 W
2	79	2.5 W
3	158	3.5 W

0,15 FLAT LEAD THICKNESS   LAND PATTERN

TYPE	A	B	C	D	E	F	G	H	I	J	K
1	6,3	3,5	3,5	1,5	1,25	0,5	8,6	2,0	3,6	6,1	2,4
2	10,9	5,9	5,3	1,4	1,5	0,5	12,8	2,0	5,7	9,3	
3	14,5	6,3	6,3	1,75	1,75	0,5	17,0	2,0	9,9	13,45	3,55

Figure 3.22 Wirewound resistors.

WATT	A	B	B₁	C	D	E	F	G	H	I	J RAD.
1	6,35	12,45	9,40	1,98	2,74	0,711	0,508	1,52	4,83	0,991	0,406
2	11,23	16,81	13,77	2,79	3,56	0,762	0,508	2,39	8,84	1,40	0,584
3	14,22	22,76	17,88	5,21	5,72	1,52	0,813	3,43	10,80	7,11	0,794
5	23,49	33,66	27,56	8,38	7,49	1,78	1,016	5,72	17,78	4,19	0,794

WATT	.05%	.1%	.25%	1%	3%	MAX VOLT
1	1.0–1K Ω	.499–1K Ω	.499–3.4K Ω	0.1–3.4K Ω	——	20V
2	1.0–2.74K Ω	.499–2.74K Ω	.499–10.4K Ω	0.1–10.4K Ω	——	54V
3				.005–2 Ω	.005–.08Ω	
5	——	——	——	.005–2 Ω	.005–.08Ω	

Figure 3.23 Wirewound precision power resistors.

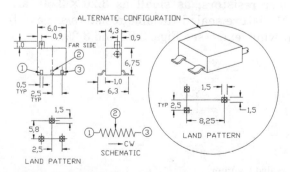

Figure 3.24 Multiturn miniature trimmer.

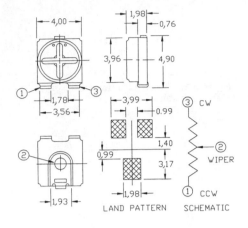

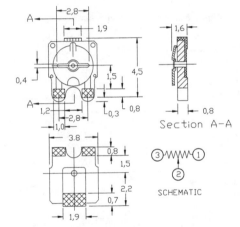

Figure 3.25 Sealed trimmer resistor.

Figure 3.26 Open-frame variable resistor.

2. Those constructed on an open frame with the resistive element exposed to the environment and solder. This type of resistor, therefore, is excluded from wave soldering (see Fig. 3.26).

Single-turn, sealed trimmer resistors, as small as 3.20×3.50, are available (see Fig. 3.27). Miniature sealed variable resistors are available with J leads and gull-wing leads (see Figs. 3.28 and 3.29).

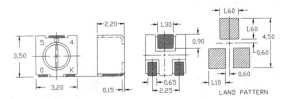

Figure 3.27   3-mm, single-turn, sealed trimmer.

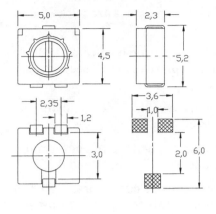

Figure 3.28 Miniature sealed J-leaded trimmer.

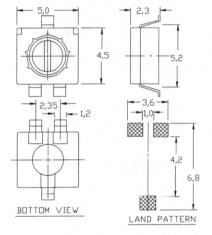

BOTTOM VIEW

LAND PATTERN

Figure 3.29 Miniature sealed gull-wing-leaded trimmer.

Resistor network.   Resistor networks, multiple resistors installed in a single package, are available with each resistor electrically independent or with all or part of the resistors interconnected. These devices are packaged in multiple leaded chip carriers (LDCCs) or as converted single in-line packages (SIPs). In SIPs the leads are alternately reformed from side to side in L shapes to permit surface mounting.

3.6.2 Capacitors

Surface mount capacitors are available in ceramic chip monolithic, tantalum, electrolytic, plastic, aluminum, and mica capacitor configurations.

Ceramic chip capacitors.   Ceramic chip capacitors are available in all standard capacitor values and voltage ratings normally used for nonpower applications (see Fig. 3.30). Chip capacitors are packaged in standard sizes as shown in Fig. 3.31.

High-voltage chip capacitors.   High-voltage chip capacitors (1000 to 4000 V dc) are available in values from 150 pF to 0.15 μF (see Fig. 3.32 for package sizes).

High-Q chip capacitors.   High-Q chip capacitors (250 V dc) are available in values from 0.1 to 1000 pF (see Fig. 3.33 for package sizes).

Tantalum chip capacitors.   Tantalum chip capacitors in conformally coated and molded packages are available (see Figs. 3.34 and 3.35).

Other types of capacitors.   Aluminum electrolytic capacitors, sealed in metal cans and installed in rectangular high-temperature plastic cases, are available in value from 0.1 to 22 μF and for working voltages of 6.3 to 63 V (see Fig. 3.36). Mica chip capacitors are available in low-capacitance values (see Fig. 3.37). Multilayer metalized polyester chip capacitors, ranging from 0.01 to 2.2 μF, are also available (see Fig. 3.38). Tube-shaped capacitors, 1.00 mm diameter × 1.60 mm long, have been developed.

Ultrathin decoupling capacitors.   Ultrathin chip capacitors, of thicknesses 0.762 and 0.16 km in the 1206 standard case size, are available as a means of conserving PWB area (see Fig. 3.39).

Variable capacitors.   Variable, film dielectric capacitors are available in open-frame configurations (see Fig. 3.40).

DIELECTRIC	CAPACITANCE RANGE	VOLTAGE RANGE	CASE SIZE
NOP/COG	0.5pf–0.1mfd	100V–200V	SEE FIG. 3.31
X7R	100pf–4.7mfd	25V–200V	SEE FIG. 3.31
Z5U/Y5V	390pf–18.0mfd	6V–12V	SEE FIG. 3.31
HIGH–Q	0.1pf–1Kpf	250V	SEE FIG. 3.33
HIGH VOLT	150pf–0.15mfd	1K–4KV	SEE FIG. 3.32

STANDARD CAPCITANCE DECADE VALUES					
1.0	1.2	1.5	1.8	2.2	2.7
3.3	3.9	4.7	5.6	6.8	8.2

Figure 3.30   Ceramic chip capacitors.

SIZE CODE	L	W	T
INCH/MM	INCH/MM	INCH/MM	INCH/MM
0403/1008	.040/1,016	.030/0,762	.030/0,762
0504/1210	.050/1,27	.040/1,02	.040/1,02
0805/2012	.080/2,00	.050/1,27	.055/1,40
1005/2512	.100/2,54	.050/1,27	.059/1,49
1210/3225	.126/3,20	.100/2,54	.059/1,49
1505/3812	.150/3,81	.050/1,27	.050/1,27
1706/4316	.170/4,32	.065/1,65	.065/1,65
1712/4432	.175/4,45	.125/3,18	.065/1,65
1808/4520	.180/4,57	.080/2,03	.065/1,65
2221/5753	.225/5,72	.210/5,33	.065/1,65
2225/5763	.225/5,72	.250/6,35	.080/2,03
2708/6820	.270/6,86	.080/2,03	.070/1,78
3409/8524	.335/8,51	.095/2,41	.070/1,78
3915/10037	.395/10,03	.145/3,68	.070/1,78
5830/14876	.585/14,86	.298/7,57	.070/1,78

Figure 3.31   Standard chip capacitor dimensions.

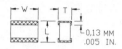

SIZE CODE INCH/MM	L	W	T
1515/3838	.150/3,81	.150/3,81	.120/3,05
1919/4848	.190/4,83	.190/4,83	.120/3,05
3333/8484	.330/8,38	.330/8,38	.120/3,05
4040/10101	.400/10,16	.400/10,16	.120/3,05
5440/137102	.540/13,72	.400/10,16	.120/3,05

Figure 3.32 High-voltage chip capacitors.

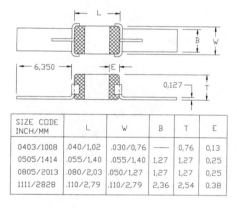

SIZE CODE INCH/MM	L	W	B	T	E
0403/1008	.040/1,02	.030/0,76	—	0,76	0,13
0505/1414	.055/1,40	.055/1,40	1,27	1,27	0,25
0805/2013	.080/2,03	.050/1,27	1,27	1,27	0,25
1111/2828	.110/2,79	.110/2,79	2,36	2,54	0,38

Figure 3.33 High-Q chip capacitors.

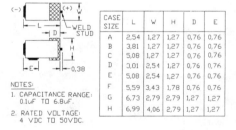

CASE SIZE	L	W	H	D	E
A	2,54	1,27	1,27	0,76	0,76
B	3,81	1,27	1,27	0,76	0,76
C	5,08	1,27	1,27	0,76	0,76
D	3,01	2,54	1,27	0,76	0,76
E	5,08	2,54	1,27	0,76	0,76
F	5,59	3,43	1,78	0,76	0,76
G	6,73	2,79	2,79	1,27	1,27
H	6,99	4,06	2,79	1,27	1,27

NOTES:
1. CAPACITANCE RANGE: 0.1uF TO 6.8uF.

2. RATED VOLTAGE: 4 VDC TO 50VDC.

Figure 3.34 Conformally coated tantalum chip capacitors.

CASE	L	W	H	T	G	R
3216	3,20	1,60	1,60	0,70	1,40	1,20
3528	3,40	2,60	1,90	0,80	1,40	1,80
6032	6,00	3,20	2,50	1,30	2,40	2,40
(5845)	5,80	4,50	3,10	1,30	2,50	3,30
7343	7,30	4,30	2,80	1,30	3,80	2,40

Figure 3.35 Molded tantalum chip capacitors.

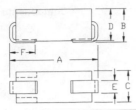

TYPE	A	B	C	D	E	F
A	9,0	4,1	3,9	3,9	2,0	2,7
B	12,2	4,1	3,9	3,9	2,0	2,7

Figure 3.36 Aluminum electrolytic capacitors.

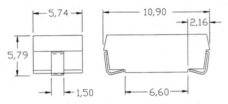

Figure 3.37 Mica chip capacitor body style.

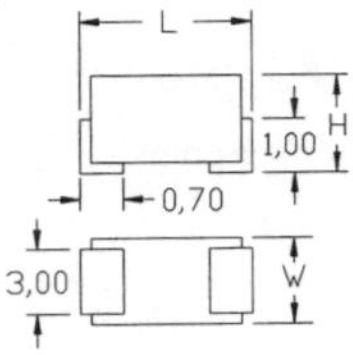

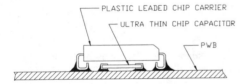

CASE CODE	L	W	H
D	4,90	4,50	2,50
E	4,90	5,50	3,00
F	7,30	5,00	3,00
G	7,30	5,50	3,00
H	7,30	6,50	3,50
L	7,30	7,00	4,00

Figure 3.38 Metalized polyester film chip capacitors.

PLASTIC LEADED CHIP CARRIER
ULTRA THIN CHIP CAPACITOR
PWB

Figure 3.39 Decoupling capacitors beneath PLCCs.

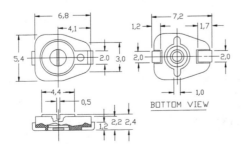

BOTTOM VIEW

Figure 3.40 Trimmer capacitor (film dielectric).

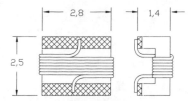

Figure 3.41 Open-frame inductors.

3.6.3 Inductors

Wirewound chip inductors. Wirewound chip inductors are available in sealed packages and open-frame construction. See Fig. 3.41 for open-frame configuration. Closed-frame packages, $3.05 \times 3.81 \times 4.57$ mm, accommodate an inductor value range of 0.01 to 1000 µH. Larger-size packages (6.35 mm) can cover a range from 1 to 200 µH and handle currents of 100 mA to 1.5 A.

Film chip inductors. Conductive film chip inductors can be made smaller than wirewound inductors and are available in values to several hundred microhenries in 0805, 1206, 1210, and 1812 standard chip package sizes.

3.6.4 Transformers

Small conventional transformers have been converted to SMT chip carriers and many configurations in open-case and closed-case packages (see Figs. 3.42 to 3.44).

3.6.5 Crystals

Crystals are available in several SMT package configurations (see Fig. 3.45 for a metal can type).

3.7 Other Components

Switches. Small surface mounted switches in J leads and gull-wing-leaded packages are available (see Figs. 3.46 and 3.47).

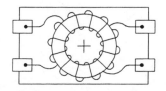
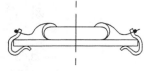

Figure 3.42 Open-frame transformer (cross section).

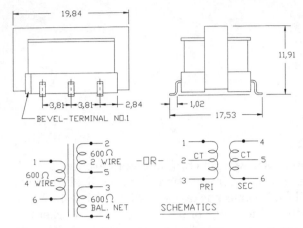

Figure 3.43 Open-frame transformers (dimensional diagrams and schematics).

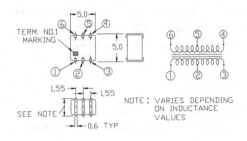

Figure 3.44 Multilayer chip transformer.

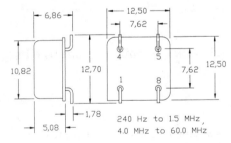

Figure 3.45 Crystal-clock oscillator.

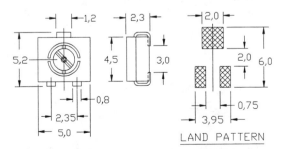

Figure 3.46 Miniature J-lead switch.

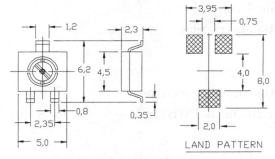

Figure 3.47 Miniature switch—gull-wing leads.

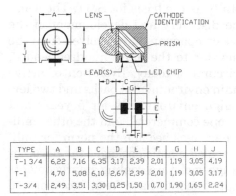

Figure 3.48 SMT PWB led.

TYPE	A	B	C	D	E	F	G	H	J
T–1 3/4	6,22	7,16	6,35	3,17	2,39	2,01	1,19	3,05	4,19
T–1	4,70	5,08	6,10	2,67	2,39	2,01	1,19	3,05	3,17
T–3/4	2,49	3,51	3,30	0,25	1,50	0,70	1,90	1,65	2,24

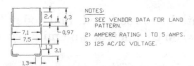

NOTES:
1) SEE VENDOR DATA FOR LAND PATTERN.
2) AMPERE RATING: 1 TO 5 AMPS.
3) 125 AC/DC VOLTAGE.

Figure 3.49 SMT fuse.

Lamps.   Small surface mounted LED lamps are available for PWB mounting (see Fig. 3.48).

Fuses.   Fuses are available in SMT packages with wrapunder, J-type leads (see Fig. 3.49).

Connectors.   SMT connectors with some or most of the following features are now available:

1. Surface attachment of contact leads
2. Very large I/O amount (≥300)
3. Low profile off of PWB

4. Attachment to double-sided assemblies

5. Zero, or low force, insertion-extraction

6. Inspectability

7. Reflow solderability

8. Ease of alignment for attachment operations

9. Positive mechanical attachment to PWB

10. Ease of cleaning

3.8 Active Components

Semiconductor devices serve as the basis for the vast majority of today's electronics industry and almost totally as the basis for SMT. The semiconductor industry began with the discovery and development of the transistor in the early 1950s. Active components have evolved from the simple, discrete, three-leaded transistor to the multi-hundred-leaded, complex, single-chip, integrated circuits of today. Evolution of active components proceeded along two main environmental paths and two levels of product life expectancy: benign environments for 5 years and rugged environments for 10 years—one commercial, and the other military. Hermetically sealed ceramic packages became the norm for military products and nonhermetically sealed plastic packages the norm for commercial products. Plastic package moisture sealing improvements and environmental endurance improvements have been incorporated during the recent past decade (see Fig. 3.50). Plastic packages now pass military environmental specifications.

The scope of commercial environments has broadened, and today there is no longer a clear distinction between commercial and military requirements (see Fig. 3.51).

Today's design challenge involves the selection of the correct combination of active component types to achieve the proper mix tailored to meet the environmental needs of each product within cost and reliability expectations.

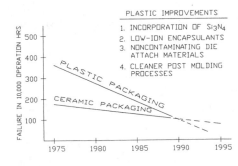

Figure 3.50 Improvements in plastic packages.

ENVIRONMENT	INDUSTRIAL SEGMENTS			
	AUTOMOTIVE	COMMERCIAL	CONSUMER	MILITARY
TEMPERATURE	-40 TO 125°C	0 TO 70°C	0 TO 40°C	-55 TO 125°C
HUMILITY	85% RH/85°C	CONTROLLED	NORMAL AMB.	85% RH/85°C
SHOCK	150g	MINIMAL	MINIMAL	1K-100Kg
VIBRATION	20g,20-2K HZ	MINIMAL	MINIMAL	20-100g/ 20-20K HZ
CHEMICAL RESISTANCE	SALT SPRAY, AUTO FLUIDS	GENERALLY, NOT RESISTANCE	NOT RESISTANCE	SALT SPRAY,

Figure 3.51 Environments per industrial segment.

3.8.1 Discrete components

Diodes. Discrete SMT diodes are packaged in leadless, cylindrical, MELF-style packages. Standard sizes are "SOD-80" (3.50 long and 1.65 in diameter) and SMI (5.00 long and 2.50 in diameter). Multiple diodes are also packaged in the multileaded, IC packages.

Transistors. Discrete transistors are encased in molded plastic packages specifically designed for non–hermetically sealed SMT application and designated as SOT-style packages for small outline transistors. Usage of this particular style of package has been expanded to include simple IC chips, dual transistors, and light-emitting diodes (LEDs).

SOT-23. The SOT-23 is the original three-leaded component developed for discrete transistors (see Fig. 3.52). Although these devices have extended leads, they offer little or no compliance to compensate for CTE mismatch other than being small and plastic. Heat flow from the chip to ambient is limited by its narrow leads.

SOT-89. The SOT-89 is larger than the SOT-23 package and more suited as power transistors and diode package, where heat transfer

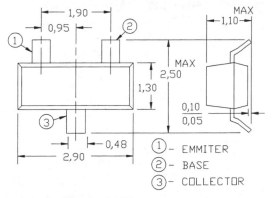

Figure 3.52 Standard SOT-23 package.

up to 0.75-W dissipation to the substrate and 1 W with heat sinks is required.

SOT-223. A newer version of SOT-23 and SOT-89 packages is the SOT-223. It features a large collector tab for more efficient heat transfer from the chip to ambient PWB. This device can dissipate 1 W on FR4 boards and 2 W on ceramic boards. It has a die size capability of up to 2.5 mm square and can accommodate diodes, thyristors, rectifiers, and MOS (metal oxide semiconductor) and bipolar transistors up to 10-GHz frequency. The larger, specifically formed lead frame can absorb a much larger thermal shock during soldering and higher CTE mismatch stress force (see Fig. 3.53).

SOT-143 and SOT-25. Four-leaded SOT-143 and five-leaded SOT-25 packages are used for components requiring extra lead(s) (see Fig. 3.54).

SOT-type packages are available in three heights for optimum processing (see Fig. 3.55).

There are no hermetically sealed SMT packages for single discrete transistors. When discrete transistors are needed for hermetically sealed products, four transistors, electrically isolated from one another, are packaged in 14- or 16-pin ceramic chip carriers. Diode bridges and other multiples of discrete active parts are also mounted in various sealed carriers.

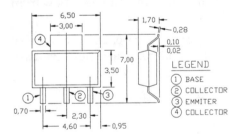

Figure 3.53 Standard SOT-223 package.

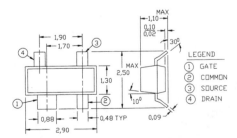

Figure 3.54 Standard SOT-143 package.

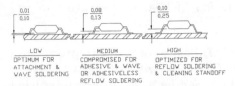

LOW — OPTIMUM FOR ATTACHMENT & WAVE SOLDERING

MEDIUM — COMPROMISED FOR ADHESIVE & WAVE OR ADHESIVELESS REFLOW SOLDERING

HIGH — OPTIMIZED FOR REFLOW SOLDERING & CLEANING STANDOFF

Figure 3.55 Standoff profile for SOT.

3.8.2 Integrated circuits

Wide variety of carrier configurations. SMT integrated circuits are available in a variety of hermetically sealed and non–hermetically sealed package types and in leaded and leadless configurations. Plastic chip carriers for non–hermetically sealed devices, ceramic carriers for long-term hermetically sealed devices, plastic leaded chip carriers (PLCCs), leaded ceramic chip carriers (LDCCs), and leadless chip carriers (LLCCs) are several of the more prominent component packaging types used in the SMT industry today.

When long-term hermeticity is not an important design requirement, plastic devices are generally selected; otherwise ceramic carriers are used. Integrated circuit carriers are either rectangular or square. Leads are positioned along two sides or around all four sides.

Standard lead pitch. Chip carrier lead pitch for SMT was initially standardized at 1.27 and 1.00 mm. As the number of leads increased, the pitch had to decrease. A newer, smaller pitch has also become a standard, 0.63 mm. This newer pitch is sometimes described as the beginning of the fine-pitch (FP) category of components and sometimes labeled as the end of the conventional, standard pitch. In the nonmetric system these pitches are known as 50-, 40-, and 25-mil pitch standards. Standard pitch sizes have not as yet been established for the FP category. If there is a standard, 0.51 mm seems to be the beginning. Other, smaller pitches range from 0.41 mm down to 0.25 mm.

Sockets. Sockets are sometimes used to mount chip carriers to PWB for special cases such as reprogrammable memory devices. Sockets are an added materiél and process expense, require extra PWB space, and generally are IMT devices requiring plated-through-hole attachment. They are available in two categories:

1. Low insertion force (LIF), which depends on a low fractional force between the socket contact and the carrier lead.

2. Zero insertion force (ZIF), where the interconnection force is applied after insertion through an actuation mechanism built into the socket or hold-down cover.

Alignment features and accommodation for heat sinking are also provided in some sockets.

SOIC packages.   Small-outline (SO) packages are the prevailing component configuration used for IC devices with 28 leads or less in nonmilitary applications. Two package widths have been standardized by JEDEC for SOICs: 3.81 mm for devices with 16 leads or less and 7.62 mm for devices with 16 to 28 leads.

"Small outline" is a misnomer when considering SMT components in general. Leads on these devices are dual-in-line and on 1.27-mm pitched centers. These devices can easily be made smaller without jeopardizing assembly fabrication yields by using all four sides. When compared to IMT DIP devices, which they replaced, they are smaller (see Fig. 3.56).

Memory devices, using the wider 7.62-mm-width SOIC package with several center pins removed, are available for ease of bus routing on the PWB. These packages are available with J leads.

SO packages with a 0.64-mm lead pitch have been developed recently. This new package, called a *shrink small-outline package* (SSOP), has maintained the 7.62-mm width and the gull-wing lead configuration but is half the length. A second modification of the SO package has also been developed specifically for memory devices called the *thin small-outline package* (TSOP). This package was developed as a means of increasing the amount of memory that could be placed in a given area.

Plastic chip carriers.   The plastic chip carrier (PLCC) is a configuration derived specifically for SMT that uses all four sides for leads with a minimum package size that accommodates the next increment of lead quantities above the SOIC package range. PLCCs can accommodate up

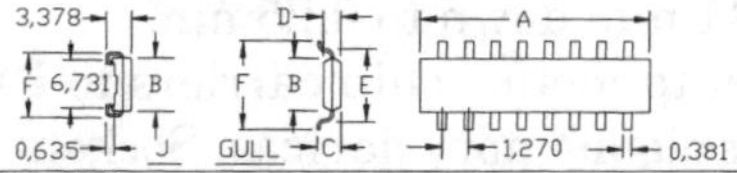

NO. OF LEADS	JEDEC STD.	A	B	C	D	E	F	LEAD STYLE
8	MD-012	4,928	3,810	1,524	1,270	5,080	6,096	GULL
14	MD-012	8,687	3,810	1,524	1,270	5,080	6,096	GULL
16	MD-012	9,931	3,810	1,524	1,270	5,080	6,096	GULL
16	MS-013	10,338	7,620	2,540	1,778	8,890	10,160	GULL
16	MD-088	10,338	7,620	3,378	2,743	—	8,433	J
18	MS-013	11,582	7,620	2,540	1,778	8,890	10,160	GULL
18	MD-088	11,582	7,620	3,378	2,743	—	8,433	J
20	MS-013	12,827	7,620	2,540	1,778	8,890	10,160	GULL
20	MD-088	12,827	7,620	3,378	2,743	—	8,433	J
20	MD-077	17,145	7,620	3,378	2,743	—	8,433	J
24	MS-013	15,418	7,620	2,540	1,778	8,890	10,160	GULL
24	MD-088	15,418	7,620	3,378	2,743	—	8,433	J
28	MS-013	17,932	7,620	2,540	1,778	8,890	10,160	GULL
28	MD-088	17,932	7,620	3,378	2,743	—	8,433	J
28	MD-059	18,136	8,890	2,540	1,778	8,890	10,160	GULL

Figure 3.56 SOs and SOJ packages.

to 84 leads [see Fig. 3.57 (flowchart in Fig. 3.57*a*; tabulated values in Fig. 3.57*b*)]. PLCC packages use J leads to fulfill two requirements: compliance for CTE mismatch and handling protection (see Fig. 3.58). These J-leaded configurations are ideal for socket mounting as well as reflow soldering directly to PWBs and can be mounted directly to leadless chip carrier land patterns without alteration.

Ceramic flat packs.   Ceramic flat packs are dual-in-line, hermetically sealed, SMT devices used with IMT assemblies before SMT became a distinctive entity (see Fig. 3.59). These packages are becoming less popular because of cost and inefficient use of PWB area.

Quad flat packs.   Quad flat packs (QFPs), initially derived for the role now served by PLCCs, is becoming the package of choice for large gull-wing-leaded devices with 0.64-mm lead pitches and lead protecting bumpers (see Figs. 3.60 and 3.61). These devices are available in plastic (PQFP) with lead counts of up to 196 pins. These devices are also available in hermetically sealed ceramic (CQFP) that can support thin-film technology and dissipate up to 30 W in the larger package sizes.

   Small metal plates have been added between the die and the bottom of the package to assist in spreading and dissipating the internally gen-

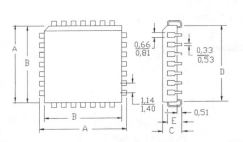

	A	B	C	D	E
20	9,78 / 10,03	8,89 / 9,04	4,19 / 4,57	7,37 / 8,38	2,29 / 3,05
28	12,32 / 12,57	11,43 / 11,63	4,19 / 4,57	9,19 / 10,92	2,29 / 3,05
32	14,86 / 15,11	13,89 / 14,05	2,79 / 3,56	12,45 / 13,46	1,65 / 3,56
44	17,40 / 17,65	16,51 / 16,66	4,19 / 4,57	14,99 / 16,00	2,29 / 3,05
52	19,94 / 20,19	19,05 / 19,20	4,19 / 5,08	17,53 / 18,54	2,29 / 3,30
68	25,05 / 25,27	24,13 / 24,33	4,19 / 5,08	22,61 / 23,62	2,29 / 3,30

Figure 3.57   J-leaded plastic chip carrier: (*a*) flowchart; (*b*) values plotted in tabular format.

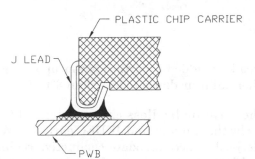

Figure 3.58   Plastic chip carrier J lead.

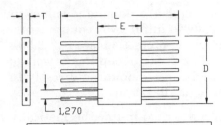

Figure 3.59 Ceramic flat pack.

NO. OF LEADS	L	D	E	T
10	24,13	6,16	6,35	1,59
14	24,13	6,16	6,35	1,59
16	24,13	10,10	6,73	1,59
20	24,13	12,83	6,73	1,59
24	24,13	15,24	9,46	1,78
28	37,78	18,03	16,19	1,78
42	33,91	26,92	16,26	2,35
48	33,97	30,79	16,32	2,35

Figure 3.60 QFP, SQFP, and VQFP standard package sizes.

PART NO. *QFP...*	PACKAGE DIMENSIONS							
	A	B	C	D	E	G	H	P
32	9,00	7,00	9,00	7,00	0,40	0,65	1,45	0,80
44	14,00	10,00	14,00	10,00	1,20	1,00	2,15	0,80
64	24,00	20,00	18,00	14,00	1,20	1,00	2,15	1,00
80	24,00	20,00	18,00	14,00	1,20	1,28	2,70	0,80
T80	24,00	20,00	18,00	14,00	1,20	1,00	2,15	0,80
✳56	12,40	10,00	12,40	10,00	0,50	1,00	2,15	0,65
✳80	16,40	14,00	16,40	14,00	0,50	1,28	2,70	0,65
✳100	24,00	20,00	18,00	14,00	1,20	1,28	2,70	0,65
✳120	31,20	28,00	31,20	28,00	0,50	1,58	3,30	0,80
✳✳48	9,00	7,00	9,00	7,00	0,50	0,65	1,45	0,50

✳ "SQFP..." ✳✳ "VQFP..."

Figure 3.61 Quad flat pack series.

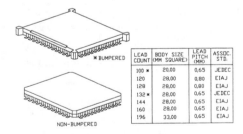

LEAD COUNT	BODY SIZE (MM SQUARE)	LEAD PITCH (MM)	ASSOC. STD.
100 ✳	20,00	0,65	JEDEC
120	28,00	0,80	EIAJ
128	28,00	0,80	EIAJ
132 ✳	28,00	0,65	JEDEC
144	28,00	0,65	EIAJ
160	28,00	0,65	EIAJ
196	33,00	0,65	EIAJ

erated heat. These heat spreaders, added to certain power application devices, have reduced the thermal impedance to less than 3°C/W.

Leadless chip carriers.   The ceramic leadless chip carrier (LLCC) package was initiated jointly by the military and JEDEC early in SMT. Four package types were derived to accommodate conductive cooling and convective, air, cooling. These packages were formulated to gain further miniaturization of hermetically sealed IC carriers. They have become a significant standard in military products.

The four-configuration standard was expanded by adding J-type leads to two of the original LLCC packages. A seventh configuration, defined as a minipak, was included in the JEDEC standard but has received little to no attention since (see Figs. 3.62 to 3.65).

PINS	LEADLESS A	B	C	D	LEADED A	B	40 MIL SERIES
16	–	–	X	–	–	–	X
20	–	–	X	–	–	–	X
24	–	–	–	–	X	–	X
28	X	X	X	X	X	X	–
32	–	–	–	–	–	–	X
40	–	–	–	–	–	–	X
44	X	X	X	X	X	X	–
48	–	–	–	–	–	–	X
52	X	X	X	X	X	X	–
64	–	–	–	–	–	–	X
68	X	X	X	X	X	X	–
84	X	X	X	X	X	X	X
96	–	–	–	–	–	–	X
100	X	X	–	X	X	X	–
124	X	X	–	X	X	X	–
156	X	X	–	X	X	X	–

X = INCLUDED IN STANDARD

Figure 3.62 JEDEC standard leadless packages.

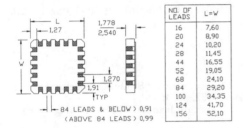

NO. OF LEADS	L=W
16	7,60
20	8,90
24	10,20
28	11,45
44	16,55
52	19,05
68	24,10
84	29,20
100	34,35
124	41,70
156	52,10

Figure 3.63 JEDEC standard leadless chip carrier.

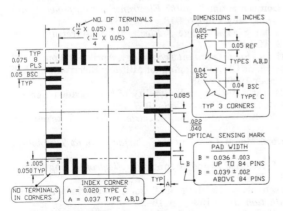

Figure 3.64 JEDEC 50-mil leadless package standard features.

PACKAGE CONFIG's.	LEAD CONFIG.	PKG. MAT'L.	CAVITY ORIENT.	COVER MAT'L.	ATTACH-MENT	PKG. SIZES
A	LEADLESS	CERAMIC	DOWN	CERAMIC	SOCKET	28,44,52 68,84,100 124,156
B	LEADLESS	CERAMIC	UP	METAL	SOCKET SOLDER	SAME AS "A"
C	LEADLESS	CERAMIC	UP	CERAMIC	SOCKET SOLDER	16,20,28 44,52,68 88
D	LEADLESS	CERAMIC	DOWN	METAL RECESS	SOCKET SOLDER	SAME AS "A"
A*	LEADED	PLASTIC	UP	TOP CAP	SOCKET SOLDER	SAME AS "A"+24
B*	LEADED	CERAMIC	UP	METAL CERAMIC	SOCKET	SAME AS "A"
MINI-PAK	LEADLESS	EPOXY-GLASS	UP	EPOXY-	SOCKET	NOT LISTED

* LEADS ADDED TO ABOVE
  LEADLESS A&B PKG's.

Figure 3.65 JEDEC basic configurations.

JEDEC and the military settled on two standard lead pitches: 1.27 mm and 1.02 mm (50- and 40-mil pitch). The military, at first, used the 1.02-mm-pitch devices almost exclusively because of the small footprint. Commercial interests stayed with the 1.27-mm devices and thereby established the 1.27-mm pitch size as the most cost-effective and most available. The military has since converted to the 1.27-mm devices. The 1.02-mm-pitch components are becoming less used and will, most likely, be phased out.

Bibliography

Anderton, John M., and Michael Sweeney: "Chip Cracking: A Study of Capacitor Failure Modes," *Surface Mount Technology,* March 1992, pp. 45–46.

Bindra, Ashok: "Tab, PQFP Packages Keep up with Advances," *Electronic Engineering Times,* September 7, 1981, p. 77.

Birzoux, M., and S. Lerose: "Plastic ICs for Military Equipments Cost Reduction Challenge and Feasibility Demonstration," *40th Electronics, Components and Technology Conference Proceedings,* May 1990, p. 918.

Buschbom, Milton L.: "Shopping for SMT Packages," *Electronic Engineering Times,* November 7, 1988, pp. T8, T19, T20.

Collins, Henry: "SMT Socket Options," *Surface Mount Technology,* April 1992, pp. 18–22.

Durtz, B: "Implications of Non-Hermetic Chip Packaging for Military Electronics," *IEPS Conference Proceedings,* September 1990, p. 848.

Engelmaier, Werner: "Surface Mount Attachment Reliability of Clip-Leaded Ceramic Chip Carriers on FR-4 Circuit Boards," *IEPS Journal,* 1987.

Hollander, Dave: "Discretes Opt for SMT and MCMs," *Electronic Times,* February 25, 1991, pp. 68–94.

———, "Trends in Discrete Surface Mount Components," *Surface Mount Technology,* May 1991, pp. 22–24.

Hutchins, Charles: "Pitting SMT Against Stand Mounting," *Electronic Engineering Times,* October 19, 1987, pp. T8–T13, T34.

Jones, W. Kinzy: "Fine Pitch Technology: Leading SMT Components into the '90s," *Electronic Packaging and Production,* June 1990, pp. 16–20.

Keister, Gerald: "Surface-Mounted Components," *Electronic Components News,* February 1990, pp. 25–39.

————, "Surface Mount Components," *Electronic Components News,* February 1991, pp. 25–44.

Kelly, Eugene: "High Power Packaging for SMDs," *Surface Mount Technology,* April 1989, pp. 43–45.

Landis, R. C.: "Plastic Chip Carriers—A Question of Reliability," *NEPCON West Proceedings,* February 1985, pp. 886–891.

Maliniak, David: "Surface-Mounted LED Lights up PC Boards," *Electronic Design,* April 11, 1991, pp. 137–138.

Prymak, John: "Customizing Multilayer Ceramic Capacitor Chips," *Electronic Engineering Times,* March 27, 1989, pp. 57–58.

Rosenfield, Maury: "Making the Move from DIPs to SOs," *Electronic Engineering Times,* November 7, 1988, p. T12.

Sack, E. A.: "Gull Wings Soar as Leads Climb," *Electronic Engineering Times,* September 12, 1988, p. 47.

Stroud, John: "Molded Surface Mount Tantalum Capacitors vs. Conformally Coated Capacitors," *Hybrid Circuit Technology,* October 1988, pp. 47–49.

Woolnough, Roger: "SMT Growth Accelerates," *Electronic Engineering Times,* November 1988, pp. T6, T18.

SMT Manufacturing

Of all the functional divisions within a company that are impacted by the introduction of SMT, none is impacted as much as the manufacturing division. Because of the extraordinary impact SMT has on manufacturing, management cannot treat SMT as just another technology in a long list of technologies and proceed to bring it into the company in a business-as-usual procedure. Introducing SMT is not just a matter of changing factory machinery. Product designs must first be changed to include SMT components and enhance manufacturability. Personnel knowledge, skill, and training need to be updated. An aggregate of new and acceptable component suppliers must be developed. Interdivisional working relationships, focused on the needs of SMT, must be nurtured.

4.1 Strategies for Implementation of SMT into Manufacturing

Implementing fully automated SMT production capability into a company that has been focused on IMT brings a life-changing impact to that company, and is best done in cautious, deliberate, and measured steps. The extent of the impact is the result of transitioning from an older, well-understood, electronic production process that is focused on the products of hand-assisted machines, to a newer production process that is focused on the products of fully automated machines. The impact is also dramatically felt because of the acute sensitivity that fully automated SMT fabrication processes have to material and fabrication variations. Insertion mounted assemblies are, to a certain extent, robust and tolerant of minor material and process variations, whereas SMT, although robust after being properly built, is very sensitive to variations during fabrication. A third reason for the scope of the impact is the effect SMT automation has on facilities, personnel, and capitalization. Factory sizes can be cut to a third, machine costs are usually doubled, and the need for technical expertise at introduction and startup can be quadrupled.

Life-changing impacts touch every functional division within the company. Design decision processes within engineering that once were the sole responsibility of engineering must now, because of the market-forced reduction in turnaround time and the need for zero defects on first pass, include participation from other divisions, particularly manufacturing. Most all manufacturing processes and procedures change. Factory sizes and layouts change. Quality acceptance and rejection criteria and evaluation methods change. Purchasing and vendor relationships change. Inventory receiving and handling changes. Personnel skills and team talent mix change. Capital expenditures can be dramatically impacted. Above all else, management involvement and interdivisional relationships, of necessity, are changed the most. To not move into SMT cautiously and deliberately is a mistake that can cause problems in every aspect of company life in the short run as well as the long run.

4.1.1   Forming an SMT strategy team

Once the company has determined its need and desire to implement SMT, the next step is to form a team of decision makers consisting of members from most of the functional operating divisions. This team would deliberate and formulate an SMT integration plan that should include (1) acquisition of SMT knowledge and resources, (2) incremental progression into SMT design and production, (3) establishment of critical dates for introduction and startup activities, and (4) determination of capital expenditures. See Chap. 2 for further discussion on an integrated SMT team.

4.1.2   Optional strategies

SMT should be introduced in a way that does not cause shock waves or sudden trauma to the company on one hand or to proceed timidly and miss company goals and opportunities on the other hand. The balance can be determined only by a well-informed, motivated team of people representing aspects of the total company and reporting directly to the highest level of management.

There are several viable scenarios from which to select a course of action.

1. *Copy the industry's beginnings.*   The industry first began in SMT by intermixing SMT chip resistors and capacitors in assemblies that otherwise had all IMT components. Next, small active SOT devices

were added to IMT assemblies, later on followed by larger SMT devices with multiple leads on 1.27- and 1.00-mm-pitch centers and mounted to mixed assemblies that became predominantly SMT. More recently even larger SMT components with many more leads on 0.63-mm and smaller pitches were incorporated on assemblies with total SMT components. In this scenario each progressive step would be taken within a full production environment.

2. *Use a full complement of component types, but begin with low quantities.* Upgrade staff and facilities gradually with exposure to total SMT component types, but at prototype and very low production levels, using manual or semiautomated machinery. Eventually evolve into medium automation and then into full automation as production orders and successful experiences dictated.

3. *Utilize outside resources.* Many companies, both small and large, begin by producing designs with the help of outside consultants and then using contract manufacturing to produce the product in the contract manufacturer's facilities. The object of this approach is to safely get into production quickly while allowing the in-house staff to "go to school" through exposure to the efforts of the contract manufacturer's experienced experts in the fully automated arena. In this way valuable training can be acquired without having to take large risks or make inordinate capital expenditures. By varying contract manufacturers, a company could also gain firsthand exposure to various machine types and process capabilities prior to making final selections themselves.

4. *Combine steps 1, 2, and 3.* A combination of the first three scenarios is the most likely approach to take to satisfy the particular needs of any one company.

4.2 Mixed Technologies

"Mixed technology" is an SMT term used in the recent past to describe assemblies with a mix of IMT and SMT components. The meaning of this term is expanding. It now includes assemblies having 100 percent SMT components that comprise a mixture of SMT subtechnologies, such as standard-pitched, fine-pitched, and chip-mounted technology (CMT) and tape-automated bonding (TAB).

It is now possible to have a mixed technology assembly with IMT components, standard-pitched SMT components, fine-pitched components, MCM devices, ultra-fine-pitched components, and CMT components all on the same single- or double-sided assembly. Wave soldering, mass reflow soldering, serialized hot-bar soldering, laser soldering, and thermal compression bonding of TAB terminations could all be required during the fabrication of that single assembly. It is feasible, but unlikely, that a single assembly would ever be that complicated. As market needs expand, however, it is probabilistic that assemblies approaching that complexity will eventually be encountered.

By mixing technologies and subtechnologies, available components can be used, some costs can be lowered, functionality can be increased, and sizes can be reduced. However, fabrication complexity would be significantly increased. In addition to the large task of developing new and challenging processes, another manufacturing challenge, for producing mixed-technology assemblies, is to get the increased fabrication steps in proper sequential order while, at the same time, keeping the quantity and quality levels high. Proper order is achieved when the number of multiple soldering steps is kept to a minimum and the total manufacturing steps are kept to an absolute minimum.

4.3   Creating, Implementing, and Managing CIM

Creating and implementing computer integrated manufacturing (CIM) is more beneficial for the larger companies, which already have established islands of automation, than for smaller companies that are semi-automated.

4.3.1   Overview

SMT, as has often been said, is not just another technology used to design and produce smaller products and be fitted into well-worn, unchanging, production procedural grooves throughout the company. By its nature, SMT, and its computer-based automation, requires a new and total company integration of people, machines, and computers. So, too, do the new quality levels, driven by market forces, require a new and totally integrated company. See Chap. 6 for the quality perspective on this theme.

To achieve the level of perfection required, the human operator should be taken out of the machine-controlled data flow loop and replaced by an automatic, computer-based system that ties all machines and other elements of the SMT assembly together to a central monitoring and control system.

Dramatic reduction in cycle time, accuracy in just-in-time (JIT) deliveries, optimization in production processes, and sales coupled to real-time production deliveries and real-time accounting are some of the improvements that should be made to compete in the new market environment. All these improvement areas have an interrelationship with one another that can be integrated into a common, computer-based, automatic data transfer system.

Through a combination of operations into an integral system, performance, costs, and quality status can be readily available in real time for management's review. The problem often encountered by management in current situations is that factories are run in real time but management data arrives on their desk in batches at intervals of time. When tied together with free-flowing data, meaningful reports and projected performances can more easily, and more accurately, be made available in real time. Investment in an integrated system could be the most important and beneficial decision that management can make, especially for larger companies.

4.3.2 Creating a CIM system

Creating a CIM system is a complex project. Organizing CIM in a relatively small factory requires the services of a dedicated team working for 6 months, and organizing CIM for larger companies can often require the dedicated services of a much larger team working for 2 years.

Tying the CIM system to current machines is generally difficult and very often requires the expertise of the machine suppliers' technical staff working closely with the CIM software vendor. First, there must be ways to input and output data from each machine over an automatic data transfer network linking the various machines to a central station. Current machines are generally equipped with magnetic tapes or floppy disks as interfaces geared to manual transfer of data. In addition, the size, density, and format of the data varies from machine to machine. The lack of data format and transfer standards within the industry has left a quagmire for would-be CAM system organizers to ferret out and compensate.

Open architecture. Open architecture is being strongly advocated to allow two-way communication between the various machines in the interest of CIM. The objective is to share information on a common database. Achieving an open architecture depends on the cooperation between machine suppliers and users.

Data is not currently on a common database. Data format is either Gerber, for photoplotting, or any one of a myriad of proprietary formats developed by individual machine suppliers. Format configurations have been developed by these suppliers to either bridge the deficiencies of Gerber or sell machines while waiting for the IPC-D-35X series to be fully developed and accepted by the whole industry as "THE" standard format. To compensate for the similar but different formats, translator software programs are currently furnished by the machine vendors and others to allow the different formats to transition into a local network.

In the meantime, maintenance of software translators and the inefficiency of data throughput, caused by translators, is costly for the industry.

4.3.3 Factory programs and CIM

Many automatic programs for individual work centers already exist. With the proper format, or translator, and a unified protocol for automatic electronic data transfer, islands of automation can be tied together into a CIM system.

```
CIM = CAD + CAM + SPC + JIT + BAR CODE + AOI + BILL/MATERIAL
      PROGRAMS + PLANNING PROGRAMS + INVENTOR PROGRAMS +
   PURCHASE/MATERIAL STATUS PROGRAMS + QUALITY HISTORY/REPORTS
             PROGRAM + ACCOUNTING PROGRAMS
```

4.3.4 CIM and operational divisions

Design engineering.   CAD data is automatically downloaded to create first-generation artwork and direct imaging, drilling, routing, component placement, automatic inspection, bare-board testing, and bills of material. CAE schematic and circuit node netlists are automatically available for testing, and thermal maps and determination of copper balance are also electronically available. Engineering could automatically receive material availability lists and cost history, process yields, and test status and SPC reports.

Manufacturing engineers.   Tool design and tool status, production flow plans, test programs, and machine control programs are automatically transmitted over the network. SPC feedback monitoring is automatically available on terminals, and malfunction reports and factory yields reported are flashed in real time.

Manufacturing.   Each workstation is equipped with a computer terminal and a bar-code reader. Terminals contain data on real-time status of products, specifically, work in progress, rework, completed work. Schedules and production plans are also available on the terminal. Equipment occupation rate and malfunction reports are automatically sent with work-around plans and new priorities and are displayed as requested. Labor and time are recorded on the terminal for each manufacturing operation. Product flow is automatically recorded on the network in real time and made available on all terminals.

Quality.   Real-time quality reports on receiving inspection, product acceptance, and discrepancies are automatically transmitted along with probable cause and current status (rework-scrap-failure analysis) of discrepant parts. SPC process and test feedback can also be transmitted over the CIM network.

Production control. Planning and scheduling factory flow, material flow, and inventory status are inserted into the CIM network. Automatic reports on product status, time and effort records, forecast completion of current work in progress, and next work to start are included on the CIM network. Bills of material are reviewed against inventory and purchase requirements keyed into CIM as necessary.

Purchasing. Materials listed by production control on the network are automatically purchased in response to schedules, needs, and packaging requirements. Status of parts (ordered, in transit, received, or troubleshooted) are automatically included in CIM.

Financial control. Personnel expenditures per job, material purchased (pending expenditures), parts received, product shipped, payments received, and performance reports are automatically tracked over the CIM network. Some financial data, under secured codes, is available on the CIM system only to designated users.

Work order. In response to sales, a work order is electronically issued over the CIM network giving all parties the authority and direction to proceed. Typically, the work order includes essential details such as customer, quantity, part number, critical schedule dates, brief description of product, and file references for further details and follows the job until completed.

4.4   Contract Electronic Manufacturing and Design Services (CEMDES)

Companies specializing in contract manufacturing represent a rapidly growing service segment within the electronics industry as a result of SMT. Once known as "assembly stuffers" for IMT, these earlier service companies consisted of rows of workbenches staffed by part-time, semi-trained labor. Now these service companies are equipped with state-of-the-art sophisticated machines and staffed by trained personnel backed by technical experts. Armed with the latest and best machines and backed by technical expertise, what these service companies now offer, that makes them so attractive, is high-quality work with very rapid turnaround time. Some of the biggest original equipment manufacturer (OEM) companies in the electronics business, such as IBM and others, employ these service companies on a regular basis.

Contract companies focused their resources, early in the evolution from IMT to SMT, on resolving the areas of SMT manufacturing where

OEMs were inefficient or inexperienced. They have expanded their value-added services to now include design, procurement of components and PWBs, burn-in and kitting, assembly, test, rework and repair, and drop shipment to primes. Some contract companies are now expanding into semiconductor fabrication and assembly services.

Contract manufacturing services ease the entrance of OEMs and others into the newer SMT technologies under controlled conditions. New technologies can be immediately introduced into products by OEMs without the risk of poor quality or misunderstood parameters. The access to contract company expertise and advanced technology factories, in the meantime, allows the internal OEM staff to gain knowledge and experience with very little risk of jeopardizing company reputation or experiencing financial loss. Capitalization of the OEM factory can also be done at a controlled rate with better-informed decisions.

Contract manufacturing services are also helpful in off-loading routine tasks that allow the contracting company to concentrate on other more strategic applications. Off-loading could also include peak production periods to allow the contracting company to maintain a balanced manufacturing operation.

In full-turnkey service, contract manufacturers buy, assemble, and test the materiél. Turnkey services involve a high up-front cash flow that only the larger contract manufacturers can provide. A few of these CEMDES contractors also provide design, but most provide only assembly and limited tests. Many contractors have recently incorporated fine-pitch assembly capability, and others are already investing in the newer MCM and CMT assembly operations.

There are drawbacks to employing the services of contract vendors. Contracting companies can lose control over their manufacturing operations to the contract manufacturers as a result of internal inactivity and the loss of in-house expertise. Financial collapse, experienced by the contractors while in the middle of a contract, is another possibility that could leave the contracting companies in very difficult circumstances.

At one time most of the contract manufacturers were operating on a shoestring. That has changed. The following is a profile of the average company in the top 20 contract companies prominent in SMT services.

1. Sales volume estimated for 1992 ranges from $35 to $220 million, averaging $42 million.

2. The number of employees ranges from 320 to 1800, averaging 745.

3. Facility area ranges from 9755 to 67,355 m^2, averaging 23,226 m^2.

4. Number of years in business ranges from 16 to 38 years, averaging 22 years.

5. Turnaround time for a prototype unit ranges from 3 to 25 working days, averaging 7 days.

6. Production turnaround time ranges from 2 to 8 weeks, averaging 4 weeks.

Overall, contract manufacturing has had a very positive effect on SMT, and as these companies experience, and propagate, leading-edge technology, they will likely continue to be a positive presence in the industry. The key to dealing with contract manufacturers is to select those with the best performance record and join with them in a long-term alliance.

4.5   Strategies in Selecting SMT Production Machinery

A universal assembly line, with a set mix of production machinery to handle any and all SMT production situations, is impractical. Each factory should be equipped with a suite of production machines that have been carefully selected to efficiently produce the intended assembly configurations at optimum throughput rates. Once selected and implemented, alternate machines could be added as necessary to produce a diverse and broader set of assembly configurations.

4.5.1   Assembly-line balance

The first and most significant consideration in organizing an assembly line is to synchronize and balance production activities all along the line. For manual assembly lines, this is done simply by adding multiples of people at any one stage until the whole line is balanced and the desired throughput is reached. Individual machine throughput capacities need to be carefully selected and balanced against all other machines along the assembly line. Some segments of an automatic SMT assembly line—that is, when lower-production component-placement machines are used—may need to be fitted with two or more machines placed in parallel with one another on the line.

Batch versus continuous-flow assembly lines.   Two basic methods are used to move products along SMT assembly lines: batch and continuous flow. In the batch method groups of assemblies are sequentially moved from one island of production process to another in remote controlled, stacked carts or by manually moved assembly containers. Continuous-flow assembly lines employ a conveyor with assemblies moving along at a steady rate throughout the complete assembly line.

Parallel assembly lines.   When single-shift throughput on one assembly line is inadequate, an entire second assembly can be added. This option is often employed for flexibility reasons. Rather than using a single line with higher-rated production machinery, this dual-line option makes the second line available for another type of assembly during times of lower production rates.

Auxiliary machines. Specialty machines, which are needed sometimes on select assemblies for their special features, can be placed on rollers and rolled on and off the assembly line as needed. Often, when machines such as high-accuracy placement machines are not needed, they can also be placed on rollers and wheeled onto the assembly as an auxiliary machine used for general assembly processing to eliminate bottlenecks.

Daily throughput. Assembly line(s) can be initially organized around a single, 8-h shift, 5-days/week work schedule ($1 \times 8 \times 5$). Work fluctuations can be handled without altering the assembly line or making additional capital expenditures by switching to any one of a wide variety of work shifts ($2 \times 8 \times 5, 3 \times 8 \times 5, 1 \times 12 \times 6, 1 \times 12 \times 7, 2 \times 12 \times 7$, etc.).

Dedicated lines. Larger factories can have one or more assembly lines dedicated to fixed production products and thereby take advantage of the cost economies that can be achieved with high-throughput but limited-configuration machines. A second type of assembly line, designed for a high-flexibility mix of assembly types, can be employed along the first to achieve a wider product mix and larger customer base and be more responsive to changing market environments.

4.5.2 Basic SMT assembly line

Typical SMT assembly lines include the following:

1. A screen-stencil printer
2. Two or more component placement machines
3. A curing station
4. A reflow soldering machine
5. A cleaning machine
6. A vision inspection module
7. A board handling system that includes a conveyor with buffer-elevators at each end of the assembly line
8. A computerized monitoring and reporting system
9. Special features that may be incorporated into one or more of the machines (which are becoming increasingly common)
   a. Bar-code readers
   b. Part presence sensor
   c. Adhesive applicators
   d. Solder paste applicator

e. Board identification

f. Integrated vision for location and orientation

g. Integrated vision for compensation and offsets

h. Computerized vision enhancement

i. Optical sensors

j. Component verifiers

k. Off-line programming

l. CAD data download

m. CAM-CIM interface

SMT can very successfully be implemented with less than full automation. Factories can be easily sized to match the production throughput rates for any company scenario. Machines are available to efficiently produce SMT assemblies in low, medium, and high quantities.

4.5.3 Component placement machines

Component placement machines are the centerpiece machines along assembly lines. Component placement is the most difficult task of the SMT assembly operations and placement machines. They are often referred to as "pick and place" machines. These are the most complex of all SMT production machines. All other equipment along the assembly line is sized to match the throughput and yields of placement machines.

Placement machine complexity range.   There are a large variety of placement machines readily available with a complexity range that starts with manually assisted simple machines having a placement capability of 1000 components per hour and available at a cost of $30,000, all the way up to very sophisticated machines capable of placing 50,000 components per hour at a cost of $750,000.

Assembly complexity, component mix, and throughput production rates are the three dominant factors considered when selecting a component placement machine.

Why use automation?   Automation should be used primarily to improve quality. The greater the amount of automation present on the assembly line, the greater is the potential for improving quality. No attempt should be made to reduce manpower per se without first improving quality. Improved quality, in time, will automatically increase production rates and yields and reduce direct labor, especially in the rework areas on the assembly lines and later in the field.

Caution when using production automation.   With the full-scale automation now made possible by SMT, problems that once could be considered as being minor must now be treated as major. An entire production lot of matériel could be irretrievably committed by the time a problem can be discovered and resolved. Design and production problems, even minor ones, must be discovered early and resolved during the design and prototype stages and verified during the initial break-in run of the assembly line.

Machine classification.   Machines are classified as follows according to their production rate capability:

1. Manually assisted

2. Low-volume

3. Medium-volume, with standard components

4. Medium-volume, high-accuracy

5. High-volume

Manually assisted machines.   Approximately 100 components can be manually placed per hour using hand tools. With vacuum pickup pencils, in lieu of tweezers, that rate can be increased to approximately 150 components per hour. With manually assisted, simple placement machines, the placement rate increases to 500 components per hour. With slightly more sophisticated manually assisted machines, the rate can approach 2000 components per hour. (*A note of caution.* Machine throughput rates are highly dependent on component types and should be viewed more for their relative value than their absolute value.)

Moderately sophisticated, manually assisted machines are simple to operate and do not require programming, and startup can be quick and training brief. With these machines, yields from manual hand-tool operations can be improved by 80 percent, operator fatigue substantially reduced, assembly boards fixtured, and components in tape and stick feeders mechanized and made more readily accessible. Machines at this capability level permit fast turnaround from one assembly type to the next, are flexible, and can be used for a wide mix of assembly configurations and, because of their relative simplicity, require minimum downtime for maintenance.

Low-volume machines.   Low-volume production placement rates are generally no greater than 3000 components per hour. These volume machines are ideal for use in entry-level operations and as backup machines along higher-production-rated assembly lines (see Fig. 4.1).

Machines in this category are still the bread-and-butter machines for many successful SMT companies. They are generally capable of placing a wide variety of board sizes and component mix within acceptable accuracies. These machines do, however, often need a constantly present

Figure 4.1 Low-volume automatic placement machine (starter system). (*Courtesy of Zevatech, Inc.*)

operator for manual loading and overseeing component feed. The purchase price for these machines can be as high as $100,000.

Medium-volume machines with standard components. Average production throughput for the medium-volume, standard-component machines is 15,000 components per hour. The basic cost for these machines can be $300,000. These machines are often referred to as "chip shooters" because they are generally dedicated to placing passive chips, MELFs, SOTs, and SOLCs supplied in tape and reel packaging. Chip shooters are not considered as stand-alone placement machines. Assembly lines that use chip shooters generally augment them with a companion machine capable of placing the larger, more complex SMT components.

Medium-volume, high-accuracy machines. Medium-volume, high-accuracy machines are the companion machines used to supplement chip shooters. These machines tend to be more complex, with computerized vision enhancement, component verification, bar-code readers, and other add-ons. Throughput for these machines, with their requirement to place more complex components at higher accuracies (i.e., for fine-pitch placement), is 10,000 components per hour (see Fig. 4.2). The cost of the basic machine can be $300,000 with an additional $200,000 for add-ons.

Figure 4.2 Fine-pitch placement system. (*Courtesy of Universal Instruments Corp.*)

Figure 4.3 High-speed placement system. (*Courtesy of Universal Instruments Corp.*)

High-volume machines. High-volume, high-speed machines can place greater than 30,000 components per hour (see Fig. 4.3). Machines dedicated to a few SMT assembly types can be capable of placing up to 50,000 components per hour. The cost of high-volume machines ranges from $500,000 to $1,000,000.

Specific capabilities of placement machines. The following list is offered as examples of specific machine capabilities that can be used as determinates in the selection of placement machines. These capabilities should match the anticipated production, the component mix, and the assembly configuration needs. Following is a list of capability categories and examples:

1. PWB size range (minimum to maximum)
2. Component types

    a. Chips: minimum to maximum sizes (e.g., 0504 to 2220, 1005 to 4532)

    b. Small-outline transistors (SOTs): 23, 89, 143, 223

    c. Small-outline integrated circuits (SOICs): 16, 20, 24, 28
    d. Plastic leaded chip carriers (PLCCs): 18 to 100 pins
3. Component feeder size and quantity
    a. 8 mm: 100 reels
    b. 12 mm: 60 reels
    c. 16 mm: 30 reels
    d. Others (24 mm, 32 mm, 44 mm, 56 mm, trays, stack tubes, etc.)
4. Placement rate
    a. Chips (no adhesive): 6000 h^{-1} (per hour)
    b. Chips (with adhesive): 2500 h^{-1}
    c. SOICs and PLCCs: 3000 h^{-1}
5. Placement tolerances
    a. Resolution: ±0.0 mm
    b. Rotational resolution: 0.1°
    c. Repeatability: ±0.02 mm
    d. Minimum part spacing: 0.04 mm
6. Modular add-on features
    a. Vision system
    b. Adhesive placement
    c. CIM interface
    d. Component verifier
    e. Dual-sided assembly automatic board flipping module (see Fig. 4.4)

On-line component storage. Component feeders on placement machines and on auxiliary carousels provide large component storage capacity on the assembly line. With extralarge, mechanized, component holding capacity, on-line storage of the entire factory inventory is possible. Line storage benefits include reduced handling and setup and reload errors.

Verifiers. Verifiers are mounted to placement machines to detect incorrect or failed components. Many companies will not start the production run if the verifiers are not on-line and ready to operate.

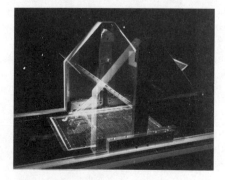

Figure 4.4 Board-flip module.

Programmable vision systems.   Competitive pressures for higher production speeds and zero defects has made visual imaging systems a necessary feature for automatic placement of fine-pitched components and fault detection. These computer-based visual systems, aided by fiducial markings and laser cameras, make it possible for the placement machine to automatically compensate in real time for pattern misalignments, bent leads, lack of coplanarity, misorientation, missing components, and other faults during production processing without sacrificing throughput. This automatic visual system is mandatory for placement of the newer fine-pitched (<0.51-mm) large-lead-count, customized components.

Vision systems have also become a necessary tool in the control of the assembly processes. Assembly variations that required compensation action on the part of the placement machine are monitored and recorded by the vision system. By feeding of this data into the SPC system, proper adjustments and corrections could be rapidly made and quality records automatically updated in real time.

Image processing, an essential subfeature of the visual system, has eliminated the need for operator intervention for many of the process variations. Parallel processing is another essential subfeature of the newer computer-based visual systems that help extend the scope of the placement machine's real-time capability. With parallel processing prodigious amounts of calculations can be made simultaneously, enabling real-time machine responses. Neural network, a computer processing methodology made possible by parallel processing, and a form of artificial intelligence used very successfully and effectively by the Department of Defense in smart weapon systems, is another high-tech (high-technology) feature now employed in vision systems to increase discrimination capability.

With the level of computer sophistication now available in automatic vision systems, it is very simple to continually upgrade the systems, through built-in model-based training techniques, matched filters, and software changes, to meet the demands of more and more complexities in SMT assemblies and meet the competition with higher and higher production rates and higher yields.

Problems that frequently confused earlier vision systems, such as similar but different shapes, variations in color and texture, variations in component body relative to lead locations, lighting and reflection variations, and the limited ability of the system to quickly process enormous amounts of calculations, are now greatly reduced by the newer vision systems.

4.5.4   Machine vendors

Machine designs, within the SMT industry, have been very responsive to the needs of user companies as forced by new technology and the demands of a changing market. Machine vendors are committed to R&D (research and development). They already have machines in the prototype stages that can handle 0.38-mm-pitched components and machines geared to 0.20-mm-pitched leads in the concept and early development stages.

4.5.5   Manual component placement

Manual placement of SMT components can be cost-effective for prototypes and very low preproduction quantities. Manual placement of SMT components, however, is much more difficult than insertion of IMT components. Additionally, quality of manually placed SMT components can be inconsistent. With IMT components, the assembler has two major advantages over SMT assemblers. IMT components are larger and easier to handle, and the PWB holes guide the leads and thereby automatically align the component properly without the operator's guidance or the need to work with visual magnification. Smallness of the components, density of the assembly, accuracy required for placing leads on solder pads, and the lack of component identification markings all combined make single-pass, zero-defect yields unlikely. In addition, large component with leads on all four sides have a unique alignment problem. Two sides can readily be aligned at the same time, but four sides presents a rotational alignment problem that gets progressively more difficult with increased size and number of leads.

Another problem with manual placement is coplanarity. Even the small-chip devices require uniform positioning in the solder paste and/or adhesive to prevent formation of nonuniform solder joints. Also the pressure used to place the component, if excessive, can cause the paste to smear, making it more likely to form solder bridges during reflow.

Manual placement requires the operator to work with plastic tweezers or vacuum picks aided by ring light magnifiers or low-power stereo microscopes ($3\times$ to $10\times$ magnification). Manual placement of SMT components, unaided by manual-assist machines, should not be attempted for even the lowest level of production.

4.5.6   Manually assisted placement

Manually assisted component placement machines require extensive operator participation. Machines that locate and deliver the proper component to the operator, either by rotating trays or bins to a window or by presenting a specific magazine, also highlight the components' positions on the assembly. It's up to the operator to pick and place the component. These machines make it possible to increase manual assembly throughput with a slight improvement in quality.

Manually assisted machines that place the components normally require the operator to find and feed the component and position the machine head over the proper PWB location. These machines have not been widely used.

4.5.7 SMT product configuration and component mix

Selection of production equipment, especially placement machines, is dictated partly by the mix of components (see Fig. 4.5) and partly by the type of assembly equipment (see Fig. 4.6). Assembly types can be any one of the following (see also Sec. 2.2.1):

1. Type I—all SMT components requiring reflow soldering only

2. Type II—a mix of SMT and IMT components with a mix of reflow and wave soldering

3. Type III—a mix of SMT and IMT components with wave soldering only

SMT assemblies can consist of a combination of any of the following component types:

1. Passive chips smaller than the head of a straight pin

2. Active components larger than a book of matches

3. Leadless and leaded components

4. Rigid-lead SOT packages

5. Fine-pitch flexible gull-wing components

6. Plastic component packages with J leads

7. Ceramic components

8. Metal packages with glass-sealed leads

9. Multichip modules

10. Chip-mounted devices

11. Insertion-mounted devices

4.5.8 Production volume and assembly mix

Selection of production equipment is predicated on the production volume as well as the mixture of assemblies. The following list contains examples of production volumes and assembly mixed that can be typically encountered:

Figure 4.5 SMT automatic placement component mix. (*Courtesy of Universal Instruments Corp.*)

Figure 4.6 Stack tubes for molded carrier ring (MCR) components. (*Courtesy of Universal Instruments Corp.*)

1. Large-volume, low-cost products produced at a steady rate

2. Low-volume, complex assemblies produced at an intermittent rate

3. Low-volume, simple products produced at a constant rate

4. High mix of assembly types, low-volume each, but high-volume total

5. Quick turnaround, high mix of fine-pitch assemblies, low production volume

6. High-volume, fine-pitch assemblies, intermittent rate

7. Larger percentage of active IMT components with few active SMT ASICs and the majority of passive devices as SMT chip components

8. Quantity ratio of passive to active 8/1, 1/1, 2/1, 1/3, and so on

4.5.9 Solder paste–adhesive application

Three solder paste–adhesive application methods are used in SMT:

1. Screen printing
2. Stenciling
3. Syringe dispensing

Screen printing and stenciling are both mass application methods for solder paste and for adhesives. Both are used for high-production quantities of standard SMT components (see Fig. 4.7). Syringe dispensing is a sequential application method.

Screen printing and stenciling.   Screen printing uses an off-contact printing method with screens mounted off of the PWB (snap-off height) of 0.50 + 1.50 mm/s. In screen printing the solder paste is squeezed through a screen pattern by a squeegee which is pulled across the screen from one side to the other at a rate of 2.5 to 3.8 cm/s with a downward pressure pushing paste ahead of it and which, by locally deflecting the screen as it travels, causes the screen to touch the PWB and deposit the paste on solder pad patterns that match the screen pattern openings. Solder paste adheres to the larger area of the pads and releases from the

Figure 4.7 Fully automatic stencil printer. (*Courtesy of HTI Engineering, Inc.*)

lesser area of the screen as the screen snaps off of the PWB. Solder paste viscosity should be 250 to 350 cP for screen printing.

With the emulsion side down, screens form a gasket that seals the opening as the squeegee passes, preventing lateral squeeze-out of paste into nonpaste areas.

Screens are made of stainless-steel mesh, mesh-nylon, or polyester woven fabric with a screen mesh count of 24 to 60 openings per linear centimeter. Mesh fibers are normally oriented at a 45° angle to the picture frame supporting the screen for standard-pitch components and 22.5° for fine-pitch components. A 0.18 ± 0.08-mm-thick film of polyvinyl emulsion is applied to the mesh and a photoimaged pattern, with opening to match the PWB solder pads, is applied. A 32-mesh stainless-steel screen with 0.08-mm emulsion will result in 0.25 mm of printed paste.

New screen can readily be made and reworked on the premises. New images (emulsion patterns) can be reworked into outdated screens. Screens should not be used with all fine-pitch geometries; the fibers interfere with the small openings. The minimum opening of 0.15 mm should be used as the lowest reasonable size of opening for screens.

Stenciling. Stencils are metal masks that are very similar to screens in shape, size, and construction (see Fig. 4.8). Both screen and stencil paste application methods use a squeegee, and both are tooled to use the same application machine. However, stencils can be constructed for direct-contact printing or, like screens, for snap action.

Snap-action stencils are made by adding screen mesh support borders between the stencil mask and the support frame. These snap-off mesh borders increased the framed stencil size by approximately one-third.

Pattern apertures are cut or chemically etched to match the PWB solder pads. Side-wall surface finish is critical. Wall finishes should be as smooth as possible without nicks, dents, dings, or rough machine tool tracks to enhance a clean separation of the paste from the stencil during the lifting or snap-off phase of the squeegee stroke. Clean walls

Figure 4.8 Fine-pitch screen printer with stencil.

are especially critical for fine-pitch applications where the PWB solder pad area should be at least two times greater than the stencil wall areas. Stencils are sometimes stepped with multiple thicknesses (stepped stencil) for assemblies that have a mix of standard-pitch and fine-pitch components. These stepped stencils make it possible to match the variable solder volume requirements and the fine-pitch area aspect ratio of pad to wall. All stencils are etched from both sides simultaneously, and the artwork is "micromodified" to compensate for the lateral etching that automatically occurs on a one-to-one ratio with vertical etching. Because of the relative complexity, stencils are generally made by specialty fabricators requiring 2 to 4 weeks' turnaround at a cost of $3,000 to $10,000 each.

Stencils have poor gasket action to prevent lateral movement of paste beneath the stencil under squeegee pressure and thus require frequent wiping of the bottom surface between prints. Contact printing works better than snap-off printing in this regard. Solder paste viscosity should be 300 to 450 cP for stencil printing.

Rather than use a stepped stencil to compensate for solder volume variations from standard-pitch components to fine-pitch components, land areas for fine-pitch leads, on the thicker standard-pitch stencils, need be only 50 percent of the pad area to achieve ideal solder volume. Staggered pattern half-size apertures are used for fine pitch in lieu of stepping local thickness down to half size.

Stencils are made of brass, which is relatively pliable, and aperture walls can be readily etched smooth and straight. Brass stencils, however, are comparatively soft and easily damaged and not compatible with all cleaning fluids and techniques.

Stainless steel, a second stencil material, is durable and compatible with all SMT cleaning methods but does not etch as smoothly as brass. Aperture walls are rough and porous following etching. Special durable coatings are used to close the pores and smooth the roughness.

Molybdenum foil is being developed as a new stencil material because it etches like brass and has the durability of stainless steel. It is, however, costly and requires the use of hazardous materials during etching.

Stencils are generally chosen over screens, even though they cost more and require twice to three times the turnaround time, because they are durable and relatively easy to clean and require much less factory floor maintenance. They can be direct-contact-printed or snap-action stencils. Snap action is made possible by adding screen mesh support between the stencil and the frame.

Polymer squeegees. Polymer squeegee blades should be sharp, square in shape, and moved at 45 to 60° angles to stencil surface. Polymer blades are normally made of 70- to 80-durometer polyurethane for screen printing and most stencil printing and 80- to 90-durometer for the higher-pressure printing of small apertures of fine-pitch stencils and high-viscosity paste.

Metal squeegees. Metal squeegee blades have recently been developed within the industry to replace the conventional polymer blades. Solder paste disposition irregularities caused by the softer polymer squeegee blades during automatic processing is the reason for the change. Squeegee blade friction is high between the polymer blade and the stencil shortening blade and stencil life. The friction pulls the stencil, thus disturbing stencil registration during printing. Polymer blades also have a tendency to locally deflect while passing over stencil apertures, resulting in a scooping action that leaves a deficiency of paste at that location. Polymer blade deflection also causes a local overpressure within the aperture that drives the paste beneath the stencil, eventually resulting in solder bridging. Blade deflection causes inconsistent deposition of solder volume, especially for the larger solder pad stencil openings. Polymer blade edges also dip into the apertures, thus reducing the amount of paste.

The new metal squeegee blades have lubricated edges and small stencil contact area, producing less pull and friction on the stencil, and allowing a uniform paste deposition that is repeatable at the higher automatic processing throughput rates.

Adhesive application. SMT components, wave-soldered to the bottom surface of mixed-technology assemblies, are adhesively bonded and cured prior to being wave-soldered. Adhesive dots are placed on the PWB in the approximate center of each component's solder pad cluster. Because of protruding IMT leads, adhesives cannot be mass-applied by screen printers or stencils, but instead, are applied sequentially with the use of an automatically controlled, pressurized adhesive dispenser located on the component placement machine (see Fig. 4.9) or a separate assembly-line station (see Fig. 4.10). Components are then placed and the adhesive cured by either heat or ultraviolet light, depending on the type of adhesive.

The adhesive application process must be carefully controlled to ensure that enough adhesive has been placed to securely hold the component through the solder wave, yet not enough to spread to the solder pads as components are pressed into the adhesive dots. Careful control is also necessary to prevent misalignment of the dots as they are dispensed. See Fig. 4.11 for a vision-controlled dispensing head, Fig. 4.12 for mechanically toggled head, and Fig. 4.13 for laser- and touch-probe-controlled heads. Figure 4.14 depicts a survival head and a dual head.

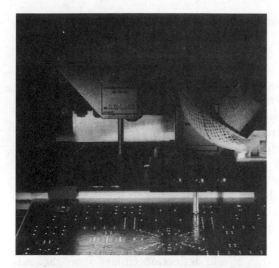

Figure 4.9 Placement machine adhesive dispensive modules. (*Courtesy of Universal Instruments Corp.*)

Figure 4.10 In-line, conveyorized, automatic dispensing station. (*Courtesy of CAM/ALOT System, Inc.*)

Adhesive dot sizes are determined by (1) the angular position of the PWB relative to the dispenser centerline, (2) the height of the dispenser needle above the PWB, (3) the nozzle size of the needle, (4) the dispensing time, (5) the type of adhesive, and (6) the pressure within the syringe. All these variables except the first two are controlled by the dispenser design. Laser-based vision measurement systems are used to automatically compensate the dispenser for the first two variables by adjusting the height and relative position of the dispenser and/or PWB to be within the customary height of one-half of the needle diameter between the nozzle and the board. Nozzle size and preheaters are used to compensate for variables in different adhesive material flow characteristics.

Dispensers.   Various mechanisms are used in dispensers to control adhesive dot sizes: timed pressure pulses, constant-pressure pump feed

Figure 4.11 Dispenser head with automation connection. (*Courtesy of CAM/ALOT System, Inc.*)

with a metering valve, constant-pressure feed with a measured piston stroke, and constant-pressure feed with a rotary screw. Dot size accuracy for the popular timed pressure pulse mechanism depends on the accuracy of the air regulator and gauge. The pump-fed metering valve requires high maintenance and the piston stroke technique, although the highest throughput requires different dispensing needles for each dot size. The technique that is accurate and repeatable and requires very little maintenance is the rotary-screw-fed dot dispenser.

For low-volume applications, small benchtop manual adhesive dispensing machines, using a timed pressure-pulse mechanism, are avail-

Figure 4.12 Dispenser head with mechanical toggle and micro adjuster. (*Courtesy of CAM / ALOT System, Inc.*)

able with a capability of placing 1000 or more dots per hour at a cost of less than $1000 (see Fig. 4.15). Prepackaged, degassed, disposable adhesive syringes are available for use with these machines. These newer syringes have resolved many of the mixing, cleaning, and atmospheric exposure problems encountered with earlier dispensing methods.

Placement machine, add-on, adhesive dispensers are available with a 5000-dot/h capability at a cost of less than $10,000, and other machines are available that can dispense 15,000 dots per hour at a cost of $25,000 (see Fig. 4.11). In-line, high-volume assembly machines are available with a magazine loader that can place 25,000 dots per hour.

Screen printing and stenciling can also be used to apply adhesive as long as the PWB is flat and there are no component lead protrusions.

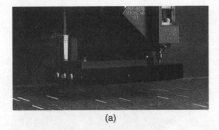

(a)

(b)

Figure 4.13 Dispenser head with (*a*) laser sensor, (*b*) touch probe programmable sensor. (*Courtesy of CAM / ALOT System, Inc.*)

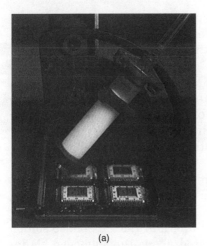

(a)

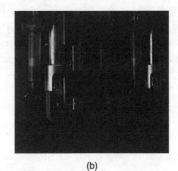

(b)

Figure 4.14 Dispenser head with (*a*) swivel adjustment, (*b*) dual head. (*Courtesy of CAM / ALOT System, Inc.*)

Figure 4.15 Automated benchtop, fluid dispense station. (*Courtesy of Asymtek.*)

These methods are mass application methods, are more consistent in applying adhesive, and are easier to automate.

Disadvantages with screens and stencils is the process development needed, and viscosity and squeegee speed and pressure are very critical. Also the PWB would require very careful handling to ensure non-smear of the adhesive.

Ultraviolet light adhesive curing is fast but often requires some preheating and, depending on the assembly and component configurations, may not be visible to the adhesive.

Solder paste dispensing. The same dispensing machines used for adhesive applications can be used for solder paste applications. Paste dispensing is becoming more common as the application method of choice for ultra-fine-pitch components.

Paste dispensing is ideal for mixed SMT-IMT assemblies and for rework and repair. Because of the high rate of stencil automatic printing, PWBs can become stacked up in front of the placement station with a queue that could result in loss of tack in the paste. Dispensing paste is also ideal when placement machine rates are between 1500 and 3000 components per hour.

Paste can be dispensed at a maximum rate of 4.4 dots per second but normally is dispensed at 2 to 3 dots per second.

Automated stenciling. Automated stenciling machines dispense solder paste from sealed paste reservoirs located above the squeegee blade. Paste remains workable in these sealed reservoirs for hours of continuous operation. Paste is dispensed only in the areas of need, and coverage is smooth and complete. Stencils are automatically cleaned on the system.

4.5.10 Postsolder cleaning

The best time to clean SMT assemblies is immediately after soldering while the assembly is still warm. Flux residue that has not yet hardened at room temperature is comparatively easy to thoroughly remove. Assemblies, therefore, should remain on the conveyor and flow from the solder station immediately into the cleaning station while the assembly is still above 50°C but less than 100°C. (See Fig. 4.16 for an entry-level cleaning system and Fig. 4.17 for a high-production cleaning station).

Cleaning SMT assemblies.   Cleaning SMT assemblies is more difficult than cleaning IMT assemblies. SMT components are mounted with less space between the component bodies and the PWB, there is less space between leads, and the center distance of large active components is further in from the component edge, all of which make cleanout beneath SMT components difficult. SMT assemblies are often very densely populated, which also makes cleaning difficult.

The twin forces of capillary action and surface tension oppose one another as fluxes and other potential contaminants attempt to flow into the tight space beneath SMT components. If contaminating fluids can readily enter the tight space, then so can cleaning fluids readily enter; the reverse is also true. The problem with reflow solder flux residue beneath the components is that the flux is deposited there prior to placement of the component and does not necessarily enter during the reflow

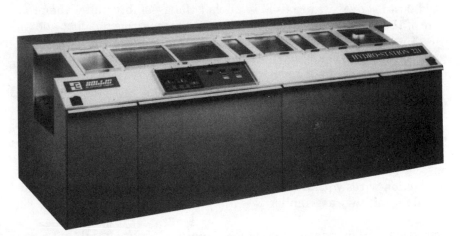

Figure 4.16 Entry-level aqueous cleaning system. (*Courtesy of Hollis Automation Company.*)

Figure 4.17 Aqueous cleaning production station. (*Courtesy of Hollis Automation Company.*)

operation. The problem with wave-solder flux occurs when there is an unplugged or nontented plated through-hole beneath the SMT component through which flux could flow upward and become entrapped beneath the component.

Cleanout of the flux residue from beneath the SMT components with tight spaces sometimes requires a combination of cleaning methods that involves pressurizing sprays, heat, fluid vibration, vapor cleaning, and/or immersion in solvents. See Chap. 6, Sec. 6.2.4 for further description of assembly cleaning.

Spray cleaning.   Conveyorized spray cleaning machines employ a series of multidirectional, forced-jet sprays pointed to penetrate the space beneath the components and to aggressively blast the open surfaces. Hold-down conveyors are used to anchor the assembly under the pressure of the cleaning spray (see Fig. 4.18). This approach is fast and allows the use of selective cleaning agents. To be most effective, spray patterns and nozzle angles should be profiled for each assembly and its corresponding cleaning agent.

Vapor cleaning.   Entire assemblies are engulfed in residue-free solvent.

Ultrasonics.   Ultrasonics vibration speeds cleaning action and allows a wider selection of cleaning agents. There is concern that the vibration associated with this approach can cause detrimental effects on internal connection wires within active devices. This cleaning method is not allowed for military assemblies.

Immersion.   Immersion can enhance cleaning. It is essential that clean solvents be used at all times with this approach, otherwise contaminated solvents may become entrapped beneath components.

Figure 4.18 Restraining conveyor system for control assembly cleaning and drying. (*Courtesy of Hollis Automation Company.*)

Heat. Heated cleaning agents work more aggressively and clean faster than do nonheated agents.

Cleaning agents. The type of flux used should be selected in consideration of cleaning. Reduced flux activity makes it possible to reduce cleaning agent activity. See "Alleged ozone depletion" subsection in Sec. 6.2.4.

The two major categories of SMT cleaning, chlorofluorocarbon-based and non-chlorofluorocarbon-based agents, include the following (the most effective agents are listed here):

Non-CFC-based cleaning agents	CFC-based cleaning agents
Terpenes	Fluorocarbon Freon-113
Isopropyl alcohol	Fluorocarbon Freon-113, alcohol blend
Deionized water	1,1,1-Trichloroethane
Detergent in (hot) water	1,1,1,-Trichloroethane, alcohol blend
Hot water	

Clean surface tests. Ionic testing uses the Omega meter to measure chlorine residue levels remaining on the assembly surface following cleaning. Ultraviolet light is used to see flux residue and particles containing trace elements incorporated in the material for postcleaning visual inspection. Surface insulation resistance (SIR) tests use an IPC test comb pattern that is probed for resistance levels above 100 MΩ at 100 V.

4.5.11 Soldering

Chapter 5 has been dedicated to soldering. See that chapter for manufacturing-related aspects of soldering. See Fig. 4.19 for midrange production wave-soldering machines and Fig. 4.20 for batch vapor-soldering machines.

4.6 Just-in-Time Inventory Control

New demands for high-mix, low-volume assemblies are reducing the need for large volume inventory and forcing manufacturers to build only what is needed, when it is needed on a just-in-time (JIT) delivery basis.

JIT theory. Just-in-time component delivery, in theory, has components being unloaded from railroad cars that have been freshly parked on the north side of a factory, and fed directly into the north-to-south assembly line with finished products being loaded into freshly parked railroad cars on the south side of the factory. Besides eliminating the cost of capital, tied up in stored inventory, hidden inventory costs are also eliminated. Hidden costs associated with stored inventory include warehousing, taxes, maintenance, insurance, deterioration of inventory (solderability), unwanted inventory, rework, and scrap. Ideal JIT systems, as theorized above, are very rare and are not practical for most companies. Even more practical JIT systems take years to implement but can be very cost-effective when implemented.[1]

JIT introduction into a company. Incorporating JIT into an existing inventory system is neither easy, nor quick, nor can it be done without risks. JIT brings about internal and external procedural changes. To be done successfully, it requires new buyer-supplier strategic alliances with long-term commitments and mutual risk taking between buyer

---

[1]A JIT case study follows. Tom Kane, vice president of manufacturing for the NCR Corporation, reported in the May 16, 1991 issue of *Electronic Business* (p. 44) that in 1980 his company averaged $1 billion in inventory to support $3.3 billion in revenue. By introducing JIT inventory control methods, Kane reports that in 1990 NCR reduced their inventory to $500 million while increasing their revenue to $6.3 billion. With JIT, NCR was able to support twice the business base with half the inventory. NCR did not have a theoretically ideal JIT system with a north-to-south assembly line; their system was less ideal but more practical. In 1980 they had a 75-day inventory supply on hand, from 400 suppliers (average per NCR factory), and by 1990 they had a 17-day supply on hand from 150 suppliers. Chrysler's Acustar, Inc., Huntsville Electronic Division, however, has a more ideal system with a normal JIT delivery window of 4 to 8 h.

Figure 4.19 Computer-controlled wave-soldering station for midrange volume producer. (*Courtesy of Corpane Industries, Inc.*)

Figure 4.20 Batch vapor-phase soldering system. (*Courtesy of Corpane Industries, Inc.*)

and supplier. Sales forecasts and general business plans, held confidentially, need to be shared between risk partners. At the same time, commitments to higher levels of quality between the parties could be agreed on.

One concern about JIT that would-be users have is the fear of being in a continual expedite mode. The opposite has been experienced by those companies involved in JIT. With strategic supplier alliances comes a smoother, more predictable flow of matériel.

Another concern, one that is more apt to occur, is the impact on inventory caused by the response of the wider, overall industry supply chain to recessionary-recovery business fluctuations. These real concerns, however, should diminish as the industry improves its manufacturing turnaround time. With more and more companies in the industry operating with less on-hand inventory, gradual fluctuations in demand will be more readily available, making finer, more responsive, control of inventory quantities possible.

4.7 Solder Application for Fine-Pitch Components

Paste solder can be successfully applied for FP component leads on 0.64-mm centers using stencil printing. Applying paste solder with a stencil mask has been accomplished for 0.51-mm-lead-pitch components but with far less success. Other methods of preapplying solder for 0.51-mm lead pitch, and below, have been developed and used for prototype quantities.

To achieve zero-defect, single-pass soldering yields with FP components the preapplied solder should be flat and uniformly thick. FP solder joints, especially those at 0.51-mm and below, are so small and so close together that there is very little margin for error. At these pitches, with so little inherent solder volume, coplanarity and predictable solder thickness become much more critical than with standard-pitch solder joints.

4.7.1 Preforms

Small, rectangular solder preforms, the size of the FP lead pad, have been automatically applied at the developmental level using prototype pick and place machines. To hold the solder preforms to the PWB (tack) and perform the customary heat transfer and oxidation removal, flux is atomized and sprayed using a spray gun through a 0.25-mm-thick stencil onto the FP component solder pads and clearance apertures for any previously soldered devices. The proper flux thickness of 0.05 to

0.08 mm must be applied uniformly across the PWB. Too little flux will not hold the preform in position, and too much flux will cause the preform to swim. By allowing the flux to dry at 25°C for 10 to 15 min before applying preforms, the flux will achieve its optimum tackiness and thereafter remain tacky for up to 3 h.

Solder preforms can be placed on the fluxed PWB using placement machines and vibratory feeder blocks with precision-located inclined slots arrayed for ease of automatic pickup. Vibration levels need to be controlled to prevent slide overlap of preforms yet strong enough to smoothly move preforms down the slope to the pickup nest at the bottom.

A second spray application of solder flux is then applied over the preforms and allowed to dry for 10 to 15 min before application of components.

4.7.2 Thick tin-lead plating

Deposition of 0.05- to 0.08-mm tin-lead plating on PWB FP leaded device solder pads can be accomplished in a two-step process. First, copper-flash all pads in the barrel of PTHs followed by 0.008 to 0.018 mm of solder plate to the same surfaces. Second, strip previous photoresist and reapply a second photoresist that covers passive component pads and exposes only the FP lead pads. Apply second layer of 0.05- to 0.08-mm-thick solder plate. Strip the second photoresist, etch the final patterns, and then reflow the solder. Although the reflow soldering—done to fuse the otherwise porous solder for oxide protection—forms a dome-shaped surface, it is still reasonably flat and usable for FP mounting. Flux coating, added during the assembly procedure, serves as the tack layer to retain the components prior to assembly reflow.

4.8 Application of LLCC Device Standoffs

One of the most critical parameters that must be controlled to achieve robust reliability of LLCC device solder joint reliability is the standoff height of the component body above the top surface of the PWB copper solder pad. For leadless devices with 32 or less leads, the standoff height should be between 0.08 and 0.15 mm, and for those leadless devices with more than 32 leads, the standoff height should be 0.18 to 0.25 mm high.

One production solution used to achieve predictable standoff heights of LLCC devices is to place and cure four uniformly spaced adhesive bumps on the PWB surface directly as rigid spacers beneath the leadless component body in the center area of the component solder pad patterns. Adhesive bump height needs to be compensated when there is either solder mash, or some other material, beneath the component body or there is solder plate on top of the solder pads.

Adhesives must be compatible with other materials that may be present, such as solder masks, and with the solder reflow and cleaning environments. Special care must be taken to be certain that the CTE and modulus of the adhesive chosen will not cause stress on the solder joints during temperature cycling of the fielded assembly.

Solder height buildup to match standoff height.   Buildup of the solder volume, to match the standoff height, can be easily accomplished by using preforms. Other methods include preapplied thick solder plate at the PWB level and solder paste applied at the assembly level in the event the solder paste does not provide the necessary solder volume on its own.

Standoffs for leaded devices.   Unlike SMT leadless devices, where standoff height is used to add compliance for CTE compensation, standoff height for leaded devices, where CTE compensation is achieved through the compliance of leads, is used to limit the height for maximal thermal transfer from the component body to the PWB. Standoff height for leaded devices can be limited in a controlled way by using thermally conductive, adhesive film with glass fiber shim cloth. Its intended use is thermal management, but it can also be used as a standoff control device for any other reason. FP leads are not as effective in dissipating heat as a corresponding leadless device. The thin leads conduct less heat than do the solder joints of leadless chip carriers. The adhesive film also serves to hold the FP device in place until it is soldered in place, and it enhances cleaning by blocking or preventing flux residue and other contaminants from getting under the component.

4.9   Evolution in Manufacturing

SMT has evolved from a combination of IMT and hybrid industries. Manufacturing environments within IMT and hybrid industries are vastly different from one another. IMT has open factories with reasonable environmental controls and personnel with generalized electronic fabrication skills. Hybrid factories, on the other hand, are dominated by clean-rooms with rigorous environmental control and personnel with special microelectronic fabrication skills. Until now, SMT products with packaged chips have been fabricated in factories converted from IMT, with relatively little impact on the facilities or personnel. However, as SMT evolves from packaged chips to unpackaged CMT chips, SMT factories are likely to change from the IMT mode to the hybrid mode. This will have a significant impact on manufacturing facilities and personnel and will require very careful planning. The transition from existing assembly processes and factory configurations to the viewer CMT factory configuration and processes, while maintaining production rates for previously contracted SMT-IMT assemblies, can be costly and difficult. What is needed is a dedicated team of experts that can plan and direct the transition without going through a dual-factory phase.

Bibliography

Balins, Paul: "CIM: The Next Step," *PC FAB,* July 1990, pp. 22–37.

Baotz, Elizabeth B., and Dave Webb: "Inventories: How Low Can They Go?" *Electronic Business,* May 6, 1991, pp. 43–48.

Biancini, John A.: "Advanced Surface Mount Design for Manufacturability," *Electronic Packaging and Production,* March 1991, pp. 40–45.

Biggs, Craig: "Soldering Techniques for Fine-Pitch SMD's," *Electronic Packaging and Production,* November 1989, pp. 66–69.

Bowers, Harry C.: "Partnering: How Contract Manufacturing Can Meet Today's Needs," *Electronic Manufacturing,* September 1990, pp. 14–15.

Cavallaro, Kenneth: "Solder Paste Dispensing versus Screen Printing," *Circuit Assembly,* October 1991, pp. 40–45.

Curtin, Mark: "Squeegee Design Affects Solder Paste Deposition," *Electronic Packaging and Production,* June 1992, pp. 76–77.

Daniels, Ron: "High-Volume Placement and Insertion Equipment," *Circuit Assembly,* February 1992, pp. 50–69.

DeCarlo, John: "Pick the Right Surface-Mount Assembly Machine," *Electronic Packaging and Production,* March 1989, pp. 72–74.

Hansolm, Scott: "Standards in Data Communication," *PC FAB,* July 1990, pp. 39–44.

Hutchins, Charles: "Screens, Stencils and Dispensers," *Surface Mount Technology,* May 1992, pp. 8–9.

Hyman, Harold: "Adding Fine-Pitch Capability to Your SMT Line," *Circuits Assembly,* November 1991, pp. 49–52.

Jean, Larry: "CIM Has Arrived," *PC FAB,* November 1991, pp. 46–50.

Lin, Andy: "Guidelines for Improving Fine-Pitch Soldering," *Circuit Assembly,* January 1992, pp. 57–62.

Lister, Pete: "Next-Generation Vision Algorithm for SMT Assembly," *Electronic Packaging and Production,* February 1992, pp. 48–50.

Maffione, Joseph: "Independents Offer OEMs Fabricating Alternatives," *Electronic Packaging and Production,* April 1992, pp. 46–49.

Mangin, Charles-Henri: "Minimizing Defects in Surface-Mount Assembly," *Electronic Packaging and Production,* October 1987, pp. 66–67.

Mankowski, Aaron, and Scott Klage: "Open Architecture in the PCB Industry," *PC FAB,* July 1990, pp. 45–53.

Nakahara, H.: "Electrodeposition of Primary Photoresists," *Electronic Packaging and Production,* February 1992, pp. 66–68.

Rhodes, Robin: "Linking the Islands of Automation," *Circuit Assembly,* October 1991, pp. 30–33.

Socolovsky, Alberto: "Demand for Contract Services is Rising Fast," *Electronic Business,* August 15, 1988, pp. 107–122.

Spitz, S. Leonard: "Midsize Manufacturers Venture into Automation," *Electronic Packaging and Production,* October 1988, pp. 70–72.

Sturgeon, Tim: "Contracting Manufacturing: A Global Picture of Supply and Demand," *Circuit Assembly (CIM Supplement),* November 1991, pp. 56–60.

Swanson, Donald E.: "A Force to Consider in SMT," *Electronic Packaging and*

Production, January 1990 (editorial), p. 7.

Walcutt, Jim: "Thermal Imaging in Fine-Pitch Technology," *Circuit Assembly,* January 1992, pp. 54–56.

Wesselmann, Carl: "Realistic Designs—Still a Problem," *Surface Mount Technology,* November 1990, pp. 65–69.

5

SMT Soldering

5.1 Soldering

The electrical and mechanical integrity of SMT interconnections depends on the integrity of their solder joints. Overall reliability of SMT is, therefore, almost solely dependent on the overall reliability of SMT soldering operations. To be successful with SMT, a company must, above all else, be successful in repeatedly and consistently producing reliable solder joints.

The range of methods available for soldering surface mounted components is surprisingly wide and varied. The opportunity for selecting the ideal soldering method that is best suited to match a particular set of production criteria is very good. Surface mounted components can be soldered by hand or with programmed machines. Solder, in the form of paste or preforms, can be applied to PWBs prior to positioning components or applied after, as in the case of wave soldering. Basic solder activation heat-transfer methods—conduction, convection, radiation, or a combination of the three—can be accomplished by using a choice of diverse heat-transfer media that include molten metal, energized light, hot gas, hot saturated vapor, radiant energy, or time-proven, intimate contact with hot soldering irons. Each of these soldering techniques has been, and continues to be, successfully used by the SMT industry. Each has also been unsuccessfully used on occasion.

Vapor-phase, infrared, laser, conduction belts and plates, convection, soldering irons, and hot bars are soldering methods most likely to be encountered in SMT.

Regardless of the soldering method selected there is one overriding, mandatory condition that must be met: the final process must be tested, evaluated, and perfected for each type of assembly and each set of production criteria before quality products can be mass-produced. Meticulous follow-up control of the process, once perfected, must thereafter be implemented whenever and wherever the process is used.

In addition to conventional wave soldering, two other soldering methods have emerged as the dominant techniques used for the majority of today's high-quantity SMT production soldering operations: (1) vapor-phase (VP) and (2) infrared (IR) reflow soldering. Infrared soldering is ideally suited for assemblies having gull-wing-leaded components with uniform distribution of components and uniformly distributed heat mass with components no closer than 5.00 mm to the edge of the board. Vapor-phase reflow soldering, on the other hand, is the method of choice for boards having a variety of leadless and leaded components, with uneven component placement and uneven thermal mass and where the component–board-edge distance can be closer than 5.00 mm.

Wave soldering is the method of choice for boards having a mixture of SMT and IMT components. Wave soldering is often the sole soldering technique used when all top-mounted components are IMT and all bottom-mounted components are wave-solderable SMT components. Combination soldering, wave-IR or wave-VP, is used when both SMT and IMT components are mounted on top and the bottom is either blank or populated with SMT components.

Hand soldering and laser soldering are often used in combination with the other soldering methods when special components and/or component terminations are not suitable for mass soldering. Some of these special components are connectors, transformers, power devices, and heat-sensitive devices.

Each of the various soldering methods is influenced, to one degree or another, by a set of variables that surrounds it. The ability to successfully manufacture SMT assemblies is, to a great extent, therefore dependent on how well the variables surrounding particular soldering methods are understood and controlled. Variables surrounding one method can be common to two or more of the other methods. Successful manufacturing of SMT assemblies, consequently, involves learning the variables and how to manage them.

5.1.1   Soldering system variables

Some of the more prominent variables that surround one or more of the soldering systems are listed below.

1. Wettability of component and PWB terminations
2. Surface contours and finishes on component and PWB terminations

3. Thermal conductivity of materials and T_g limitations
4. Component lead pitch
5. Component lead configuration(s)
6. Component lead orientations
7. Land patterns on PWBs
8. Component placement density
9. Thermal and cleaning sensitivity of components
10. The mix and placement on SMT and IMT components
11. Thermal density of components and PWB materials
12. Presence of thermal interconnects on components and PWBs
13. Presence of constraining layers in PWBs
14. Thickness of PWBs
15. PWB size and material type
16. Presence of adhesives
17. Plated through-hole size and aspect ratio
18. Presence of solder masks
19. Ability to clean all areas and surfaces
20. Composition of solder alloys
21. Composition of flux
22. Need for off-line soldering
23. Rework and repairability
24. Production quantity and throughput
25. Product life-cycle reliability
26. Cost goals

Each soldering method also has its own set of process characteristics and limitations that sometimes require alterations of the flux and/or solder paste compositions. Assembly configurations also have their own set of particular characteristics that may require special tailoring of processes.

One of the most successful methods of promoting reliable soldering is to ensure that the temperature of metal surfaces involved in the solder joint has reached equilibrium by the time reflow begins (see Table 5.1 and Fig. 5.1). Metals, however, have two thermodynamic characteristics, thermal conductivity and thermal capacitance, that quantitatively differ from one metal to the other and, consequently, complicate the

TABLE 5.1   Time (in Minutes) to Reach Temperature (−55 to +125°C)

Material	Cold to hot	Hot to cold
96% Alumina	2.0	3.8
Epoxy glass	3.0	4.1
Polyimide	2.5	4.0

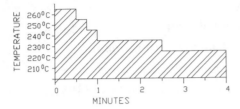

Figure 5.1 Reflow time and temperature maximums.

way temperature equilibrium is reached. Precise tailoring of the time-and-temperature profile of the soldering cycle for each assembly is the control method industry uses to overcome these variables.

5.1.2 Soldering profiles

Rate-of-temperature change, dwell times, and temperature levels are the three elements typically used to profile the ramp-up, equalization, reflow, and cool-down phases of the soldering cycle (see Figs. 5.2 and 5.3).

The following list is a brief description of the chemical, metallurgical, and thermal dynamics occurring during the solder cycle phases:

1. *Ramp-up.* As a general rule, the soldering cycle begins by raising the assembly temperature to 110 ± 10°C at a rate not to exceed 3°C/s, although 4 to 5°C/s has been used successfully under some circumstances. Within these temperature boundaries, and with the use of IR lamp heaters, a safe, rapid evaporation of solder paste volatiles can be accomplished. Without the use of IR lamps as the source of heat during this initial period, an earlier, separate, bake-out operation may be needed to remove the solder paste volatiles. Because of the rapid heating during this initial period, thermal gradients within the assembly are developed. By momentarily stopping the temperature rise at 110°C, the unevenly distributed thermal energy accumulated to that point is given time to equalize, thus greatly reducing the possibility of causing detrimental thermal shock to sensitive components at the next-higher temperature level. During this delay in the ramp-up period, flux activators, within the solder paste, begin scrubbing and reducing oxides from metal surfaces.

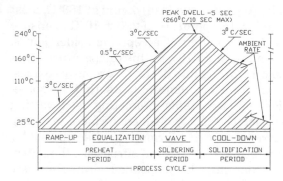

Figure 5.2   Wave-soldering process cycle.

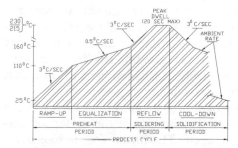

Figure 5.3   Reflow soldering process cycle.

2. *Equalization.* Assembly temperatures, now nearly equalized, are once again raised, only this time at the much slower rate of 0.5°C/s until a temperature of 160 ± 10°C has been reached. This period can be thought of as a "soak" period. It is during this period that materials with slower heat-transfer characteristics are allowed to catch up with the material having faster transfer characteristics so that these materials do not experience delamination or other types of thermally caused shock damage. Most of the remaining solder paste solvents are driven off. Fluxes are fully activated, cleaning termination pads and component lead, and promoting surface wetting. This soak period is generally accomplished in less than 1 min.

3. *Reflow.* It is generally considered safe to rapidly raise the temperature of PWBs and components from 160°C to the 220°C reflow level: within this relatively narrow 60°C temperature rise, thermal differences within the materials and components have proved to be nondestructive. Therefore, in the interest of throughput, minimization of component exposure to elevated temperature, and the minimization of any further formation of copper-tin intermetallics, the rate of thermal change is once again set at the more rapid pace of 3°C/s. Although the

melting point of eutectic solder (63% tin/37% lead) occurs at 183°C, most users operate with a solder reflow temperature of 220 ± 10°C to ensure quick and total reflow and the formation of acceptable solder joints. Typically total reflow occurs in less than 10 s at this temperature level. Just prior to this period, and continuing into it, the flux removes oxides, that are always there, and other surface contaminants, that are sometimes there, such as sulfides and carbonates, promoting good wetting and creating a flux blanket over the newly cleaned surfaces just long enough to prevent the recurrence of oxide formation while the solder wets and holds. As the temperature reaches, and surpasses, the eutectic melting point, individual solder particles suspended within the paste begin to melt and quickly run together, forming a molten mass that flows toward the hotter spots, pushing the flux away from the cleaned surfaces, wetting and covering them until all contiguous surfaces are covered with solder and the competing forces (i.e., surface tension, metallurgical-chemical attractions, capillary action, component buoyancy and gravity) have reached a state of equilibrium and the final joint configuration has been formed. It is precisely at this moment in time that the next step, the cool-down period, must begin.

4. *Cool-down.* The initial phase of this last period involves a rapid cool-down that is just as critical to the reliability of the solder joint as the earlier portions of the soldering cycle. The initial rate of cool-down has a dramatic impact on the solder grain structure, with a faster cool-down forming a tighter, smaller, better fatigue-inhibiting grain structure. Fast cool-down also puts a stop to the copper-tin intermetallic growth. The cool-down rate should be neither faster nor slower than the rapid elevation rate of 3°C/s. This cool-down rate should be maintained until the assembly reaches 160°C. Thereafter the cool-down should be in accordance with a natural descent within the room environment.

5.1.3 Cautionary note

The soldering profile suggested above is ideal for components and materials in general but could be detrimental to PWBs with a low T_g level. A tradeoff between the need to preheat components, soak time, temperature levels, and damage to low-T_g (glass transition) PWBs must be conducted prior to committing the final design and manufacturing fabrication processes. It may be necessary to use a higher-T_g PWB or to subject components to higher ramp-up temperature deltas. Here, experience and development are necessary.

5.2 Major SMT Soldering Methods

Wave, infrared, and vapor phase are the three major SMT soldering methods used for high-production mass soldering.

5.2.1 Wave soldering

Wave soldering is a conductive heat-transfer system in which molten, tin-lead solder is used as the medium to transfer heat to the workpiece assembly as well as being the source and means of applying solder to the assembly. The assembly travels along on a conveyor at a predetermined rate of speed through a carefully controlled wave of liquid solder, which forms the electrical and, for surface mounted components, the mechanical interconnections between the components and PWB.

The earlier, single-wave, version of this soldering method was, and still is, the only method available for mass production soldering of IMT assemblies. A second wave was added to the basic, single-wave system to accommodate Type II and III SMT assemblies. The directional flow for the single wave in the original version had two optional configurations: unidirectional or bidirectional (see Fig. 5.4). Both configurations had relatively gentle and smooth wave motions. Unlike the original wave, the newly added SMT wave, located in position 1, is agitated and intentionally vigorous to improve wetting and promote complete solder flow around the tightly packed SMT components, especially in the shadow areas behind the small, rectangularly shaped, bottom-mounted chip components as they plough through the solder wave (see Figs. 5.5 and 5.6). The original wave, in position 2, remains gentle and smooth.

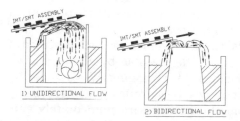

Figure 5.4 Original wave-soldering flow options.

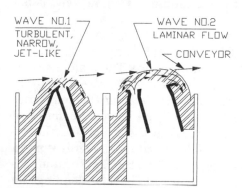

Figure 5.5 SMT dual-wave soldering.

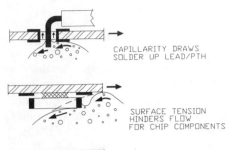

CAPILLARITY DRAWS
SOLDER UP LEAD/PTH

SURFACE TENSION
HINDERS FLOW
FOR CHIP COMPONENTS

SURFACE TENSION
HINDERS FLOW FOR
LEADED COMPONENTS

Figure 5.6 Wave action with IMT-SMT soldering.

A typical wave-soldering system includes stations that first apply flux to the bottom surface of the assembly and up into the plated through-holes, a second station that preheats the assembly, and a third station that applies the solder.

Wave soldering is a proven technique for IMT assemblies and is ideally suited for high-throughput production of mixed-technology applications involving Type II and III styles of SMT assemblies. Machines, facilities, and much of the know-how exists.

In the past oil was added to the wave's solder pots to serve as a blanket and as a scavenger inhibiting the buildup of solder dross (the oxides of tin and lead). Oil was also added to reduce solder surface tension, to improve wetting, to add a coating to the board against oxidation, and to make it easier to remove flux. With the closer component spacings of SMT, the added process advantages made possible by adding the oil can be all the more fruitful in reducing solder bridging. Wave soldering can also, like all other soldering methods, be significantly improved by eliminating the air (oxygen) from the atmosphere immediately surrounding the assembly during soldering operations. Nitrogen is generally used as the oxygen-free gas replacing the air. Like other soldering processes, the tolerance level for the maximum oxygen allowed is between 5 and 10 ppm.

Three benefits gained by replacing air with nitrogen are (1) uniform solder joint appearances for both SMT and IMT components, (2) less dross formation, and (3) less machine maintenance. The major benefit, however, is indirectly linked to compliance with the Montreal Protocol, an international agreement between nations limiting the industrial use of alleged ozone depletion chemicals, and the alleged depletion of the earth's ozone layer. The nitrogen reduces the oxidation; the reduced oxidation, in turn, makes it possible to reduce the activators in the flux. With lower levels of activators in the flux, the flux residues become less tenacious, thereby making it more likely that cleaning the residue with cleaning fluids other than the black-listed, chlorofluoro-carbon-based fluids can be accomplished. With nitrogen it may be possible to eliminate cleaning altogether. A smaller amount of activators in the flux also reduces the flux charring and board discoloration at the worst-case, high-temperature level. With reduced oxidation, the possibility of using bare copper PWBs becomes more likely.

There is one major reason why wave soldering will continue to be used with SMT: it is the only mass soldering technique that can feasibly be used to produce mass soldering joints for IMT components. Because of their relative position to the wave, IMT components also experience lower temperatures during wave soldering than do the totally immersed SMT components during IR and VP soldering.

Wave-soldering usage of SMT is, however, limited. Except for the passive chip components and a few other devices, wave soldering cannot be used for SMT components. For the SMT devices that can be wave-soldered, they must first be attached with adhesive, an added operation not necessary with the other soldering methods. The other wave-soldering disadvantages for SMT assemblies are the fluxing of the entire PWB surface, compounding cleaning complexities, and the strict board layout requirements for component orientation and spacing that inhibits miniaturization.

5.2.2 Infrared soldering

Infrared (IR) reflow soldering is a noncontact, medium-free, heat-transfer system that uses radiant energy as its primary heating method. Unlike other soldering methods, which heat objects from the outside surface inward, IR is capable of heating objects from the inside outward as well as from the outside inward, thereby raising the temperature of objects with suitable emissivity characteristics more rapidly yet with less internal thermal gradient difference.

IR soldering overview. IR heat sources operating in nonvacuum environments heat the gas medium in which they are operating, causing the intended object to be heated by a dual source of heat energy—radiant and convection. The percentage of heat from one source or the other can be varied by changing the type of IR emitter, adding secondary emission panels, forcing the atmospheric gas against the object while isolating the emitter, using baffles and shields, or changing full-spectrum IR panels for tungsten-quartz lamps.

The IR heating process is sensitive to materials, component densities, and dark colors with the danger of overheating one component and underheating another. Each assembly needs to be thoroughly scrutinized and its IR sensitivity signature profile established prior to IR soldering.

Primary control of the soldering process is gained by subdividing the IR heating chamber into individually regulated zones. Heat output levels from the emitter, in any particular zone, can be altered by adjusting the input voltage and exposure time. Shields can be added to block heat from reaching particular components and baffles placed to redirect forced gas. The heating tunnel can be made longer for added zones. Transportation through the sequential zones in the heating tunnel is provided by a belt conveyor. Altering the speed of the conveyor is another method of controlling the heat profile. With very efficient operations, conveyor speeds can reach 3 m/min. Each assembly can be custom-profiled using these techniques.

Process results of IR soldering can also be altered by changing the atmospheric gas within the heating chamber. Air is the most plentiful gas and the least expensive, but it is not oxygen-free. Pure nitrogen is oxygen-free and readily available and, with the exclusion of oxygen from the reflow chamber, can reduce throughput time by 20 percent. By adding a small amount of hydrogen to the nitrogen the tendency for flux to spread can be reduced.

IR reflow soldering. IR reflow soldering, at first, could be used for only a limited number of assembly configurations, and then only after carefully profiling the machine zones with baffles, reflectors, and shields to protect assembly vulnerability against the otherwise excessive heat concentration typical of earlier IR systems. Recent improvements in IR source heaters and convection heating have broadened the number of assembly configurations that can successfully be reflow-soldered by IR.

IR radiant-heat-transfer mechanism. IR radiant heat is thermally transparent to most media (air, gas, glass window, etc.), and heats SMT assembly materials that are in the direct line of sight of the IR energy waves in accordance with the material's characteristic response to the particular IR wavelength emitted by the source heater. The air or gas temperature, however, remains neutral. Line-of-sight IR heating of SMT assemblies has a severe process problem. Solder joints that happen to be in the "shadow" of other components experience heat starvation and results in those starved joints becoming faulty.

Four material responses to IR radiation. Surfaces of SMT assembly materials thermally respond to IR radiant heat energy waves in one of four ways, depending on the emissivity characteristic of the material and the wavelength of the incoming energy:

1. *Reflected.* All the IR energy is reflected away from the surface without heating the surface (example: mirrors).

2. *Transparent.* All the IR energy freely passes through the surface and material without inducing heat (example: window pane).

3. *Opaque.* All the IR energy is absorbed by the surface and turns to heat (example: tar roofing).

4. *Semitransparent.* Some of the energy, depending on wavelength, freely passes through and the remainder of the energy is absorbed and turns to heat. Semitransparent responses occur in nonhomogeneous materials or composites.

IR frequency bands. Traditionally, infrared electromagnetic wavelengths have been subdivided into the following three frequency bands because of distinctive response characteristics of materials:

1. Near IR (shortest wavelengths)

2. Middle IR

3. Far IR (longest wavelengths)

Thermal response of materials used in SMT assemblies fall into two general frequency bands, popularly known as (1) shortwave IR (near-to-middle wavelengths) and (2) longwave IR (middle-to-far wavelengths).

IR tungsten lamps, operated at peak power, produce shortwave IR radiant heat. IR tungsten lamps, operated at less efficient lower power, and IR diffused panels, produce longwave IR radiant heat.

Convection heating mechanism. Comparatively large reflector panels, typically ceramic, are heated either by IR lamps operated at lower power or by resistance elements generating longwave IR energy. These reflectors, with their relatively large heated panel areas, in addition to becoming secondary emitters in the longwave range, heat the air (or gas), which, in turn, heats the SMT assemblies. The hot air, in this way, heats the SMT assemblies, in combination with the IR radiated energy, by convection heat transfer.

Some characteristic material responses to IR

1. Thin organics are largely transparent to shortwave IR; thicker organics are less so.

2. FR4 printed wiring boards, which are composites of different materials, react as semitransparent materials and heat from the inside out as well as outside in.

3. Chip packages, which are opaque and comparatively poor thermal conductors, heat slowly and consequently can be temperature-profiled to reduce thermal shock.

4. Solder paste is opaque to longwave IR and semitransparent to short-wave. When heated by shortwave IR, paste heats from the inside out as well as the outside in and thereby more gently drives out the otherwise more volatile solvents.

5. Component leads are somewhat reflective to shortwave energy, which can have a direct impact on solder wicking.

IR tungsten lamps.   Tungsten IR lamps consist of a thin-gauge tungsten filament housed in an argon gas-filled, hermetically sealed envelope of vitreous quartz. The IR power (and frequency) emitted from the lamp's surface is proportional to the voltage applied. Operational control, within IR reflow equipment, is generally performed by turning the lamps on and off in response to temperature sensors. Source voltage remains at a set point.

IR diffused panels.   One face of a relatively large panel of low mass refectory insulation, in a metal housing, is layered with a resistive foil element covered by a glass-ceramic secondary emitter layer. It is through these glass-ceramic surfaces that IR energy is emitted.

Classification of IR solder reflow systems.   IR solder reflow systems have evolved over the years from simple, single-source heating devices to today's sophisticated mix and balance of source heaters that heat primarily with radiated shortwave, longwave, or fullwave IR energy combined with convective heat transfer, nitrogen or air atmospheres, and forced air (or gas) and judiciously placed baffles and reflectors.

Class I IR system. When high-temperature, tungsten filament lamps, emitting shortwave IR energy, are incorporated as the heat source in a conveyorized tunnel oven for reflow soldering, the soldering equipment is categorized as a Class I IR system.

Class II IR system. When IR lamps are operated at lower, less efficient temperatures, emitting longwave IR energy instead of the customary shortwave, and combined with ceramic reflector panels that are radiation-heated by the lamps and heat the atmosphere surrounding the SMT assembly, supplementing the IR radiated heating with convective heating, the soldering equipment is categorized as a Class II IR system.

Class III IR system. When IR emitter panels are used, which emit radiated heating energy in the longwave IR range in combination with reflectors, the equipment is classified as a Class III IR system. Convection heating within Class III equipment can be naturally induced by the draw of slow-moving hot air or gas through the oven chamber and out the vents. Convection forces can also be controlled by the selective use of blowers or compressed air.

Class IV IR system. The use of plenum chambers makes it possible to block the line of sight and reflected waves of IR radiated energy from direct contact with assemblies and by using forced-air/gas flow the assemblies can be heated primarily by convection forces. Reflow equipment operated primarily with convection are categorized as Class IV IR systems.

Design features that enhance IR reflow soldering. The following design features will alleviate some major heat distribution problems for IR reflow soldering systems:

1. Components should be dispersed in a uniform pattern and not in randomly dispersed clusters of bunched components.

2. Large components should be placed near the corner or edges of assemblies to position them in the areas that tend to be hottest within IR conveyorized machines.

3. Metal planes, judicially designed into the PWB along the edges and corners of the top and bottom surfaces, can serve as energy reflectors and heat conductors to reduce accumulation of heat in the hot areas.

5.2.3 Vapor-phase soldering

Vapor-phase soldering is the term applied to a condensation-soldering technique that is ideally suited to reflow solder on surface mounted component joints. In-process SMT assemblies with solder paste and components in place are immersed into a hot, saturated, oxygen-free vapor which condenses onto all exposed surfaces of the assembly releasing its latent heat to the assembly, producing a temperature high enough to rapidly reflow the solder (see Fig. 5.7).

At the heart of this soldering system is an oxygen-free, saturated vapor generated under atmospheric pressure by heating a unique fluid formulated to boil at exactly 215°C (other exact temperatures are also available). Since it is a continuous distillation process, the vapor condensate, which engulfs and heats the assembly, is inherently clean for each assembly. Vapor phase also produces rapid, even heating of the entire assembly regardless of geometry or thermal mass.

Introduced by The 3M Company as one of their "Fluorinert" family of electronic liquids, the liquid used in vapor phase is chemically inert,

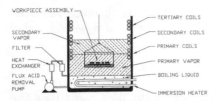

Figure 5.7 Batch vapor-phase apparatus.

nonflammable, and nontoxic; has high thermal stability and a high dielectric strength; is compatible with plastics; dries without leaving a residue; is nonpolar; and is nonhazardous to the ozone layer. It is also colorless and odorless, has a low boiling point relative to its molecular weight, and has low surface tension. Because of the ease of setup and prove-out and imperviousness to part geometries, vapor phase is the system of choice for short production runs of highly complex SMT assemblies having the widest range of board geometries and component configurations. Because vapor phase produces evenly distributed temperatures over unevenly distributed thermal masses, it is ideal for high-volume production of assemblies that consists of PWBs with large, laminated power and ground planes and large quantities of ceramic components.

Because vapor-phase soldering occurs in an oxygen-free environment, there is less need for active flux, and, consequently, a less aggressive, postsoldering cleaning can be utilized. Vapor phase does, however, require a higher capital investment, higher maintenance costs, and more stringent vigilance in monitoring and controlling the facilities. The basic system requires the following subsystems:

1. An immersion heater to boil the primary liquid creating the working vapor

2. A sump-pump subsystem to filter the flux that continuously washes into the boiling sump during condensation, a sump pump, a counterflow heat exchanger, a filtration unit, and a dryer

3. An exhaust treatment system to carry away the toxic thermal-chemical decomposition products

4. A water cooling subsystem that includes condensing coils, water chiller, pump, and a surge tank

5. A continuously recirculating secondary vapor injection system (for a batch type of vapor-phase system) located above the primary vapor keeping the primary vapor from escaping into the air; a subsystem

that includes a chemical reservoir, injectors, chemical and mechanical filters, a pump, and a water separator

Vapor phase has several functional disadvantages:

1. Pure vapor-phase systems relentlessly and rapidly heat everything to the one reflow temperature, producing thermal shock damage in some components.

2. Vapor phase can also produce open solder joints because of its inherent tendency to heat component leads before heating the matching PWB pads and thus causing an inordinate amount of solder to wick up the leads, before other balancing forces within the solder cycle can take effect, thereby depleting the volume of solder away from the critical lead-to-pad interface.

3. Solvents remaining in the paste after the bake-out can explode, causing components to shift and solder balls to form.

4. Longer dwell times at the reflow temperature level promote growth of detrimental amounts of copper-tin intermetallics.

5. In-line machines that are not horizontally compensated for inclined conveyor travel into and out of the vapor zone, or do not include vibration dampers can, along with gravity and the return flow of condensate fluids, cause the components to shift to a tilted and/or tombstone position during the critical cool-down phase.

Vapor-phase soldering machines can be altered to permit a more graceful ramp-up of assembly temperatures by adding IR preheaters with profile control (see Fig. 5.8). Combining the temperature ramp-up advantages or IR with the consistent, uniform temperature of vapor phase eliminates many of the problems associated with vapor phase, and extends the applicability of vapor-phase soldering to an even wider range of product types.

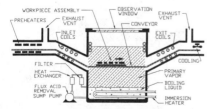

Figure 5.8 In-line vapor-phase machine.

5.3   Minor SMT Soldering Methods

Conductive, laser, hot-bar, hot-gas, and hand soldering are the minor SMT soldering methods used for prototypes, rework, and limited production.

5.3.1   Conductive soldering

In conductive soldering heat must first be conducted through the PWB and then into the joint to be soldered. The basic heating element used for this soldering method is a hot plate. A single, stand-alone hot plate is often used in small prototype facilities. In-process assemblies with preapplied solder paste and preplaced components are positioned on the hot plate long enough for the heat to be transferred through the board and reflow the solder joints on the opposite side, at which time the assembly is removed and allowed to be cooled naturally or with forced air. A series of hot plates adjusted to sequentially increasing temperatures are positioned in line with a conveyor belt transferring the in-process assemblies from one plate to the other in a timed sequence to allow preheating of the assembly and then the reflow of the solder joints. The conveyor then moves the reflowed assembly through a cool-down phase.

Conductive soldering is the earliest method used in SMT and is still used. These machines are relatively simple and inexpensive and can fit within small manufacturing areas (see Figs. 5.9 and 5.10).

Conductive heating does raise the temperature of the PWB to the solder reflow temperatures. It is limited to single-sided assemblies only and dependent on flat PWBs for effective heating.

5.3.2   Laser soldering

Lasers are used for specialty soldering today, but they could be used for general production in the future. As lead pitches for larger percentages

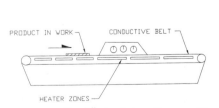

Figure 5.9 Conductive belt reflow machine.

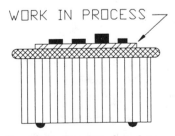

Figure 5.10 Benchtop hot plate.

of components drop below 0.63 mm and the thermal and mechanical fragilities of these components rise, most of the mass soldering techniques used today will not work. As frail leads become even more frail, and as the need for closer coplanarity between the lead and PWBs goes beyond the capability of nonmachined surfacing techniques, it will become necessary to treat each solder joint as a separate entity. Automatic machinery with on-line processing and the ability to judge good from bad in real time, and with corrective action capabilities, will be needed. Some laser machines are already being improved by adding smart subsystems which bring discernment and judgment capabilities on-line.

Laser soldering provides inherent capabilities that are unique to laser systems alone and unmatched by other soldering methods:

1. Lasers can deliver controlled amounts of heat with pinpoint accuracy to individual solder joints that have leads with diameters as small as 0.03 mm without raising the temperature of the parent component or adjacent components.

2. Components and leads do not move during the noncontact soldering and cool-down cycles.

3. Solder reflow is accomplished within 100 to 800 ms, eliminating the threat of intermetallic growth altogether.

4. Laser soldering is relatively free of solder balling or bridging, removing the need for solder masks.

5. Laser soldering is not affected by SMT assemblies with metal core inner layers and heat sinks.

6. Accuracy and ultrafast soldering leave multilayer PWBs thermally unaffected, eliminating the need for preheating.

7. Special atmospheres are not necessary to achieve high-reliability soldering, nor are special chemicals required.

8. Toxic fumes are not produced by the process.

9. Adhesives need not be high-temperature types, normally needed for other soldering systems, to survive reflow or wave temperatures.

Some newer, smart machines have infrared feedback temperature measurement sensors that monitor and map the laser solder joint temperature profile during the reflow cycle and communicate results to an on-board minicomputer which, in turn, alters the time duration of directed laser energy to compensate for detected joint variations. The computer also collects the joint data and makes it available for statistical process controls and other quality auditing.

Laser soldering has one major disadvantage. It can solder only one joint at a time and, consequently, is unlikely to replace current mass soldering techniques for production of standard-size SMT components. Lasers also require special safety procedures on the assembly line and are more expensive to purchase, and redevelopment of processes and retraining of employees are also more difficult.

5.3.3 Hot-bar soldering

Hot-bar soldering, as the name implies, is a conductive reflow soldering technique that uses a 310 $\pm$ 30°C bar under pressure to simultaneously solder all the leads along one side of a multileaded SMT component (see Fig. 5.11). Hot bars are either arranged in pairs to solder dual-leaded components or arranged as quads to solder leads on four-sided components. This technique is becoming prominent as a production soldering method for fine-pitch QFPs (quad flat packs) and TAB (tape automatic bonding) devices.

The critical parameters associated with this soldering technique are the temperature profile, the pressure profile, and the coplanarity between leads and PWBs.

Hot-bar soldering, like laser soldering, is a sequential process and not a mass process and therefore throughput is limited. It has, however, proved to be a successful production method for lead pitches and sizes below the standard 0.127 and 0.063 mm.

5.3.4 Hot-air–hot-gas soldering

Hot-air and hot-gas (not to be confused with convection and IR) are used almost interchangeably. This method is more or less limited to individual component soldering. The hot air (or gas) is generated by a benchtop machine that blows air (gas) through a nozzle that encompasses all the solder joints (27) on a given component. Individual nozzles are required for each component size. This soldering technique is used primarily for rework and repair but can also be used for production. The machines are small, simple, and inexpensive.

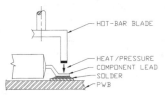

Figure 5.11 Hot-bar solder joint.

5.3.5 Hand soldering

"Hand soldering" is a popular term used to describe manual soldering, but not soldering iron usage only. It is not unusual for assemblies to have manually applied solder paste and manually placed components and then be mass-reflow-soldered using a batch reflow system.

Hand soldering of SMT assemblies has in the recent past been generally limited to prototypes, touchups and repairs, odd-shaped components, power devices, interconnection devices, and thermally sensitive components. Now, hand soldering is widely used in the attachment of fine-pitched components.

Soldering irons, hot plate, hot bars, stand-alone vapor-phase vats, IR, and hot-air ovens are devices used for manually oriented soldering operations.

Soldering iron. It is imperative that operators be specifically trained, certified, and annually recertified for SMT manual soldering. The best senior operators, not novices, should be selected for SMT manual soldering. Skilled IMT operators, however, cannot be expected to produce qualified SMT solder joints without first being retrained specifically for surface mounted assemblies. SMT tools, timing, and temperatures all differ from those used in IMT. Hidden ground planes and constraining layers within multilayer boards present a much larger thermal load than those experienced in earlier technologies. Small surface pads used in SMT PWBs have far less peel strength. Newer materials and high-density terminations that restrict tool size and challenge operator dexterity are additional difficulties and differences that are encountered in SMT manual soldering.

Operators especially need to learn solder timing and solder quantities. They need to know what good SMT solder joints look like and what bad solder joints look like for the various leaded and leadless configurations. Operators should also be trained on realistic soldering tasks. Rejected assemblies are often saved and used for this purpose.

Unlike IMT components, which are held in place by their clinched leads, SMT components must be held in place by external means at least until one terminal or pin is soldered. Tweezers, tape, adhesive, and temporary holding compounds are some of the holding methods normally used.

Small-diameter wire solder (0.6 to 0.8 mm in diameter) is ideally suited for hand attachment of most SMT components. Conventional soldering irons are suitable for a few SMT applications but unsuitable for most others.

Using a high-wattage soldering iron with a massive tip capable of a high rate of heat transfer is one way of increasing SMT production throughput; it is more often a way of damaging components and scorching PWBs. Soldering irons and tips should be properly matched to the components and PWB for the SMT assembly at hand. Soldering irons should have the capability of repeatedly generating and delivering sufficient amounts of heat directly to the solder joints and then automatically recovering each time quickly enough to repeat solder at a rate of 20 to 25 joints per minute. For soldering irons to be safe and have the proper amount of heat control, a self-regulating temperature control system needs to be included in the iron itself. For flexibility and rapid maintenance, the tips should be readily replaceable. The iron must not produce electromagnetic interference and should be suitable for usage in electrostatic discharge protected areas. Preset temperatures of soldering irons should be kept in control within 5°C and have a maximum temperature range between 300 and 400°C. The weight, size, and balance of the soldering iron should also permit close, tedious work without affecting the endurance or diminishing the dexterity of SMT operators during a normal workday.

5.4 Solder Paste

Solder paste has become an indispensable part of SMT production. The vast majority of surface mounted components are now attached with solder paste. Solder paste is used in both high- and low-rate production. With the exception of assemblies designed for wave soldering, all other assemblies with component lead pitches of 0.63 mm and above are designed for paste soldering. The ability for mass placement of small, controlled amounts of solder and flux at specific locations and the paste tackiness that holds components make solder paste ideal for SMT assembly application.

Solder paste is made of small, spherically shaped particles of solder that are uniformly suspended within a mucilage composed of flux and various thixotropic agents. Solder particles are graded by size and issued in size groupings (see Fig. 5.12). The mucilage serves as the carrier vehicle for the solder, controlling rheology and tackiness. Solder, flux, and agents are combined into a homogeneous paste having the consistency and flow characteristics of chilled butter. Combinations of these ingredients are formulated by the paste supplier to meet specific and standard needs of the users.

Solder paste is commonly considered as a two-part mixture of metal (the collection of solder particles) and organics (the conglomerate of flux and other constituents). The weight-to-volume ratio between these two parts has a significant impact on paste shelf life, viscosity, and application methods and the volume of the final reflowed solder joint (see Table 5.2).

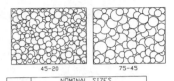

| TYPE | NOMINAL SIZES | | | NOTES: |
	>1% (LARGE)	80% (MIDDLE)	20% MAX (SMALL)	
1	150	150-75	75	1. 25 TO 45 (-325 +500 MESH)
2	75	75-45	45	2. 45 TO 75 (-200 +325 MESH)
3	45	45-20	20	3. ALL DIMENSIONS IN MICRONS
4	38	38-20	20	

Figure 5.12 Solder paste particle size and distribution.

TABLE 5.2 Solder Paste Application Viscosities

Application	Viscosity, cP
Pin-dot transfer	50–250
Syringe dispense	200–450
Screen print	500–700
Stencil print	600–900
Fine-pitch stencil	≥1000

Typically, the solder accounts for 85 to 92 percent of the paste weight and only 40 to 55 percent of its volume (see Fig. 5.13). With this disproportionate weight difference the solder tends to settle toward the bottom of the paste container during prolonged shelf life: paste in this separated condition is unusable. If all other ingredients are homogeneously distributed and the chemical properties of the organics are unaltered, then the paste may be rejuvenated by stirring. If the paste is dry and the solder is settled, then discard the container of paste.

The most commonly used solder alloys for SMT are 63 Sn/37 Pb (tin/lead) and 62 Sn/36 Pb/2 Ag (tin/lead/silver). The eutectic alloy, 63 Sn/37 Pb, is most widely used because of its singularly unique feature, when heated to reflow temperatures, of going directly from solid to liquid, bypassing a plastic range (see Fig. 5.14 and Tables 5.3 and 5.4).

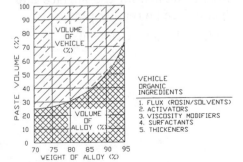

VEHICLE ORGANIC INGREDIENTS

1. FLUX (ROSIN/SOLVENTS)
2. ACTIVATORS
3. VISCOSITY MODIFIERS
4. SURFACTANTS
5. THICKENERS

Figure 5.13 Volume/weight ratios of alloys and vehicle.

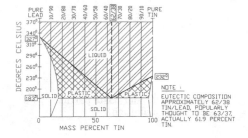

Figure 5.14 Tin-lead phase diagram.

TABLE 5.3   Melt-Reflow Temperature of Prominent SMT Solder Alloys

Alloy	Melting temperature, °C	Reflow temperature, °C
63 Sn/37 Pb	183	208–223
60 Sn/40 Pb	183–190	210–220
62 Sn/36 Pb/2 Ag	179	204–219
10 Sn/90 Pb	268–302	322–332
5 Sn/95 Pb	305–312	332–342

TABLE 5.4   Flux Decomposition Temperatures

Flux type	Decomposition temperature, °C
Rosin	273
RMA flux	275
Synthetic resin	375

5.5 Flux

The primary function of flux is to remove oxides and other mild contaminants from the mating faces of conductors and the surface of solder, making it possible to solder the conductors together while in the unaltered environment of the earth's atmosphere (see Fig. 5.15). Specifically, flux performs the following functions when heated:

1. Chemically reacts with oxides, lifting them from the surface and forming soluble compounds which can then be readily removed

2. Protects the surface it just cleaned from being reoxidized

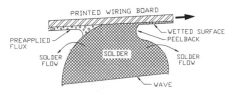

Figure 5.15 Flux displacement in wave soldering.

3. Participates in transfer and distribution of heat

4. Enables solder to wet the cleaned surfaces

5. Reduces the surface tension of solder, allowing it to spread freely

6. For paste solders, serves as the carrier and vehicle for solder particles, organic and inorganic solvents, rheology modifiers such as waxes and oils, activators, and other additions

7. Serves as the vehicle for adjustment of solder paste rheology

Water White Rosin, the gum from pine trees, has served almost exclusively as the soldering flux for the electronics industry. In its natural form it is inactive at ambient temperatures, but when heated to soldering temperatures it melts very rapidly, causing the abietic acid, inherent in the natural rosin, to react with copper oxide, forming soluble copper salts for ease of removal.

Rosin is a solid at ambient temperatures that can be crushed and reduced for use in a solvent. Rosin is nonsoluble in water and is an excellent insulator. Supplemental ingredients, such as amines, amine hydrochlorides, or organic acids, are added to the rosin to increase its chemical activity, thereby speeding up the soldering process and extending its operational range to include less solderable workpieces. Natural rosin, with its relatively benign reaction, serves as the base for a solder-flux grading system that describes the activity levels of rosin fluxes when supplemented by additives:

1. Type R—nonactivated (natural rosin)

2. Type RMA—mildly activated (rosin with additives)

3. Type RA—activated (rosin with further additives)

4. Type Synthetic Resin—nonremoved residue

5. Type OA—water-soluble

6. Type SA—synthetic activated, solvent-soluble

The last three flux categories (see items 4 to 6) are non-rosin-based fluxes recently developed and incorporated into production operations in response to alleged local and global environmental needs.

Synthetic resin fluxes have low solids content and result in noncorrosive residue that does not interfere with probe testing and need not be removed. Water-soluble fluxes are equivalent to RMA fluxes in activity and use a variety of activator types that leave a corrosive residue that *must* be thoroughly removed. These flux residues can be removed by aqueous cleaning with clean water (an inordinate amount is generally needed) without detergents. Synthetic activated fluxes, like water-soluble fluxes, are very active and leave residues that must also be removed. Unlike water-soluble fluxes, synthetic activated fluxes should not be used with vapor-phase soldering. Water-soluble fluxes are not soluble in the primary fluorinated working fluids of vapor phase and consequently can easily be filtered out through the normal processes. Synthetic activated fluxes, however, are soluble in fluorinated liquids (but not in water) and can result in deposits and coking, from the synthetic activator breakdown and decomposition products, building up on the surface of the heaters submerged within the boiling fluorinated fluids, and subsequently releasing toxic products (see Table 5.4).

In the past, rosin-based flux residues have been effectively removed by using chlorofluorocarbon (CFC)-based solvent cleaning. These solvents will soon be banned because of their alleged contribution to the depletion of the ozone layer. The search for a replacement, a replacement that is as effective a cleaning fluid as CFCs, has been given priority within the electronics industry.

Fluxes used with SMT soldering are applied either separately from the solder, as with wave soldering (see Fig. 5.16), or as an integral part of solder paste, as with IR and VP soldering. When used separately, fluxes can be applied by dipping, brushing, spraying, wave fluxing or foaming. Foaming fluxes are produced by using liquid flux forced by compressed air through porous ceramic cylinders that generate fine, evenly sized bubbles which burst on contact with the PWB. Foam fluxing is generally limited to fluxes having a solids content below 35%. Wave fluxing is preferred for dense assemblies with narrow spacing, fine lines, and multilayer boards with plated through-holes. Wave fluxing is used for fluxes with higher amounts of solids (up to 60%). Dip flux-

Figure 5.16 In-line automatic spray fluxing system.

METAL TO BE SOLDERED	SOLDER-ABILITY	ROSIN FLUXES			ORGANIC WATER SOLUABLE	INORGAN. WATER SOLUABLE
		NON-ACT.	MILD ACT.	FULL ACT.		
GOLD COPPER TIN SOLDER SILVER	EASY TO SOLDER	YES	YES	YES	YES	NOT RECOMMENDED FOR ELECTRICAL SOLDERING
NICKEL CADMIUM BRASS LEAD BRONZE	LESS EASY TO SOLDER	N/A	N/A	YES	YES	YES
KOVAR	DIFFICULT	N/A	N/A	N/A	YES	YES
ZINC STEEL CHROMIUM MONEL	VERY DIFFICULT TO SOLDER	N/A	N/A	N/A	N/A	YES

Figure 5.17 Metal solderability and flux selection.

ing is simple and inexpensive, but lacks uniformity and cannot easily be put under statistically process control. Brush fluxing uses rotating bristle brushes in a foaming flux head. Brush fluxing can be used for manual operations involving limited production and large-size PWB and component variations. This process also lacks fine control. Spray fluxing requires adequate ventilation and, being a potential fire hazard, requires safety precautions. Spray fluxing results in excess flux on the trailing edge of the board and requires that measurements be taken of content ratios at regular intervals to compensate for evaporation of flux solvents. However, with the introduction of low-solids fluxes, enclosed spray systems, and enclosed pressurized flux reservoirs, spray fluxing is becoming a prominent system for SMT assemblies, especially those with fine-pitch devices. With these improvements spray fluxing offers the best means for uniform and consistent deposition of flux and is compatible with high- and low-volume production.

Flux effectivity on various metals is shown in Fig. 5.17.

5.6   Solder Preforms

Solder preforms are premanufactured, solid, solder shapes formed in precise amounts to fit specific solder joint configurations. This method of soldering is often used to form high-reliability joints for large leadless components mounted higher off the board, for fine-pitch-leaded devices, as substitutes for solder paste in applications where the reflow process method has not been perfected, and for solder attachment of special components. Preforms are held in place by the tackiness of the flux which must be applied prior to placement of the preforms. Preforms can be placed manually or with the same pick-and-place machine that places the components.

Preforms can be manufactured to tight tolerances in sizes as small as

0.25 mm. They are shaped either for individual solder joints or as continuous strips of solder foils or ribbons. They are available in a large variety of shapes and alloys and can also be custom-ordered. Packaging is required to protect the preforms from physical damage and oxidation. Any of the SMT reflow soldering methods are usable with preforms.

Bibliography

Anjard, Dr. Ronald P., Sr.: "Solder Paste Factors Affecting SMT Quality and Reliability," *Printed Circuit Assembly,* June 1987, pp. 20–23.

Bernard, Bob: "Full Spectrum Infrared Energy for Surface Mount Attachment," *Hybrid Circuit Technology,* September 1987, pp. 39–42.

Buddle, Klaus, Birger Panzer, and Ferdinand Quella: "Avoid the Toxic Danger of Vapor-Phase Soldering," *Electronic Packaging and Production,* December 1988, pp. 68–71.

Butherus, Ted F.: "Effects of Infrared Soldering on Thermoplastics," *Surface Mount Technology,* February 1989, pp. 11–13.

Capillo, Carmen: *Surface Mount Technology, Materials, Processes, and Equipment,* McGraw-Hill, New York, 1990, pp. 207–249.

Coombs, Clyde F., Jr.: *Printed Circuits Handbook,* 3d ed., McGraw-Hill, New York, 1988, pp. 23.3–23.29.

Cox, Norman R.: "Near IR Reflow Soldering of Surface Mounted Devices," *Surface Mount Technology,* October 1986, pp. 27–30.

Dow, Stephen J.: "An Examination of Convection/Infrared SM Reflow Soldering," *Surface Mount Technology,* April 1987, pp. 13–15.

Dow, Stephen J.: "The Future of SMT Soldering," *Surface Mount Technology,* January 1991, pp. 54–57.

Down, William H.: "Achieving Consistency in IR Reflow," *Electronic Packaging and Production,* January 1987, pp. 60–63.

Elliott, Donald A.: "Q&A: Nitrogen Wave Soldering," *Circuits Assembly,* October 1991, pp. 55–59.

Flattery, David: "Infrared Reflow Solder Attachment of Surface Mounted Devices," *Connection Technology,* February 1986, pp. 24–29.

Harper, Charles H.: *Electronic Packaging and Interconnection Handbook,* McGraw-Hill, New York, 1991, pp. 5.1–5.57.

Hendrickson, Hal: "No-Clean Wave Soldering: The Challenge of the '90s," *Surface Mount Technology,* August 1991, pp. 34–36.

Hodson, Timothy L.: "Applied Technology—Spray Flexing for Today's Soldering Processes," *Electronic Packaging and Production,* January 1992, p. 47.

Hutchins, Charles L.: "Optimizing the Vapor Phase and IR Reflow Processes," *Electronic Packaging and Production,* February 1988, pp. 106–108.

———, "SMT/FPT Soldering Problems and Solutions," *Surface Mount Technology,* July 1990, pp. 37–40.

Hwang, Jennie S.: "The Future of SMT Soldering," *Electronic Packaging and Production,* November 1990, p. 63.

Hyman, H., and D. J. Peck: "Vapor Phase Soldering with Perfluorinated Inert Fluids," *Proceedings of the Technical Program NEPCON,* 1979.

IPC-SP-819: *General Requirements for Electronic Grade Solder Paste,* Institute for Interconnecting and Packaging Electronic Circuits.

Keeler, Robert: "Specialty Solders Outshine Tin/Lead in Problem Areas," *Electronic Packaging and Production,* July 1987, pp. 45–47.

———, "Lasers for High-Reliability Soldering," *Electronic Packaging and Production,* October 1987, pp. 29–35.

———, "Preventing Reflow Solder Defects," *Electronic Packaging and Production,* February 1988, pp. 98–100.

———, "Soldering Iron for Production," *Electronic Packaging and Production,* January 1990, pp. 44–49.

————, "Selecting Soldering Iron Tips," *Electronic Packaging and Production,* December 1991, pp. 26–31.

Linman, Dale L.: "Optimizing Vapor Phase Reflow Soldering," *Electronic Packaging and Production,* November 1990, pp. 43–47.

Markstein, Howard W.: "Vapor Phase Soldering," *Electronic Packaging and Production,* July 1982, pp. 33–44.

————, "SMT Reflow: IR or Vapor Phase?" *Electronic Packaging and Production,* January 1987, pp. 60–63.

Marshall, James L., and Steven R. Walters: "Fatigue of Solders," *The International Journal for Hybrid Microelectronics,* 1987 (1st quarter), pp. 11–17.

Maxwell, John: "Temperature Profiles. The Key to Surface Mount Assembly Process Control," *Surface Mount Technology,* July 1990, pp. 22–26.

Mellul, Benoit Thote Syvie, Benedict Maire-Freysz, and Dominique Navarro: "Remove Air and Boost Quality of IR Reflow Solder Joints," *Electronic Packaging and Production,* July 1991, pp. 76–80.

Miller, Carl B.: "Lasers as Reflow Soldering Tools," *Hybrid Circuit Technology,* July 1988, pp. 48–51.

Mims, Forrest M., III: "Hand Soldering SMC's," *Radio Electronics,* November 1987, pp. 71, 72, 87.

Miurs, Miroshi: "Overview of YAG Laser Soldering Systems," *Electronic Packaging and Production,* February 1988, pp. 43–49.

Nayer, H. S., and S. M. Adams: "Reflow Soldering with Reactive Gases," *Electronic Packaging and Production,* November 1990, pp. 39–40.

Olsen, Dennis R., and Keith G. Spanjer: "Improved Cost Effectiveness and Product Reliability through Solder Alloy Development," *Solid State Technology,* September 1981, pp. 121–126.

Peans, Stefan: "Past Reflow Soldering," *Surface Mount Technology,* April 1992, pp. 39–41.

Rieley, Dick: "Making Solder Joints by Laser," *Electronic Packaging and Production,* November 1990, pp. 48–51.

Rubin, Wallace: "An Inside Look at Flux Formulations," *Electronic Packaging and Production,* May 1987, pp. 70–72.

Saille, M., Vlaeminck, R., and D. Avau: "Copper Multilayer Structures: An Electrical and Analytical Evaluation (Thick Film Paste)," *Hybrid Circuit Technology,* April 1986, pp. 37–41.

Samsami, Darius: "SMT Reflow: Facing the Challenges," *Electronic Packaging and Production,* January 1991, pp. 65–67.

————, "Engineer's Fact File: Solder Paste," *Electronic Packaging and Production,* March 1991, pp. 35–36.

Schuette, Richard: "Selecting a Laser Job Shop to Fit Your Requirements," *Hybrid Circuit Technology,* October 1988, pp. 9–11.

Slattery, James A., and Paul A. Socka: "A Quick Guide to Solder Preforms," *Electronic Packaging and Production,* June 1991, pp. 79, 80.

"Soldering System Increases Efficiency at Motorola," *Electronic Packaging and Production,* February 1990, pp. 146–148.

Stein, Michael Alan, and Stephen J. Muckett: "Solder Paste for Reliable Surface Mount Assembly," *Hybrid Circuit Technology,* September 1988, pp. 8–12.

Taylor, Barry E., Joel Slutsky, and John R. Larry: "Technology of Electronic Grade Solder Pastes," *Solid State Technology,* September 1981, pp. 127–138.

Wassink, R. J. Klein: *Soldering in Electronics,* Electromechanical Publications Limited, 1984.

Williams, David J.: "Thermoplastic Compounds for IR Solder Compatible Interconnect Devices," *Connector Technology,* May 1991, pp. 37–41.

Zarrow, Phil: "IR Reflow Soldering Systems and Steps," Circuit Manufacturing, 1990.

————, "Optimizing IR Reflow," *Electronic Packaging and Production,* November 1990, pp. 32–37.

Quality Assurance

6.1 Market Effect on Quality—An Overview

A worldwide, quality paradigm seems to be occurring in which products with mediocre workmanship and early malfunction are no longer being tolerated; for this reason, industrial stakes are much, much higher today for electronics manufacturers. Mean-time-between-failures (MTBF) standards, which were once tolerated, are now unacceptably low. Customers at all levels of the economy are apparently saying that they will no longer accept products that "just get by" because equivalent but superior products are readily available at competitive prices. Customers are demanding better reliability and getting it. They have also stopped buying products that have a reputation for being mediocre. All of this is having a profound impact on the electronics industry. All segments of the industry, and their infrastructure supply chains, from the individual component part supplier to the completed assembly supplier, are affected. Companies who are committed to improving their own quality level are very reluctant to buy from companies who do not share that same commitment, regardless of the supply company's size or prominence. It seems likely that, in time, no company will be immune from the impact of this quality paradigm.

6.1.1 Worldwide initiatives

Customer demands are being turned into worldwide quality standards and causing governments to initiate ways of stimulating and propelling their national industries into pursuing higher levels of quality. Product quality is no longer a matter of national pride; it is becoming a matter of national economic survival. The U.S. government has established the Malcolm Baldrige Award, Japan established the Deming Award, and now the European Quality Award has emerged. The International Organization of Standardization (ISO) has developed a quality system offering certification and guideline standards (ISO 9000 to 9004 series). The aim of ISO is to help companies set up procedures for achieving a high level of quality in products and services and to instill confidence in customers that the certified company can deliver quality products. ISO 9000 has been favorably received and, being ISO 9000 certified, is becoming one of the essential pass-keys to doing business in Europe and, from all indications, soon in the United States.

6.1.2 Impact on companies

The impact of this change in quality on the industry as a whole is profound; for some individual companies this sudden change in customer expectations and demands is better described as an upheaval.

To achieve the level of quality that the new demands make, responsibility for quality can no longer rest solely on a select staff of personnel who examine end products just prior to shipping. The change that is necessitated by this "upheaval" in the company—if the company intends to respond, and therefore survive—needs to be deep and needs to be cultural.

Response within the company is deep because it affects all personnel: from the board of directors declaring the company's commitment to achieve the higher quality levels, and their own affirmative personal commitments, to the shipping clerk packaging the end product.

The change is also cultural because the responsibility for company quality is no longer limited to one group whose primary job was to inspect the end product. Now, quality begins at the beginning and is consciously pursued by every involved person all along the production path to the delivery of the final goods or services.

Perfection and organizational excellence have to become the two main goals ingrained in the soul of the company. They must become a corporate way of life at all levels. Otherwise, the level of quality now demanded—a level far above current practices—will never be met.

Quality as a descriptor has gained a much wider connotation; it goes beyond the hardware, and now includes services and the customer's total satisfaction. To meet these new demands, management's commitments have to be refocused from profits as the dominant, daily conscious preoccupation to commitments to achieve total customer satisfaction. In the long run profits still must be the overall results of company activities, but profits can more naturally be achieved when the customer is totally satisfied.

Customer satisfaction, as implied by this new quality, extends beyond the product; it includes all dealings as well as all services. Customers at every level of the economy are becoming to expect perfection. For a company to pursue anything less than perfection would be a mistake.

6.1.3 Impact on SMT

SMT also requires total company commitment to perfection. Achieving maximum SMT manufacturing yields also begins at the beginning. SMT and this new total company quality endeavor are a perfect match; one complements the other and together they work synergistically.

Because of total automation and the two-axis surface mount orientation in SMT, SMT production is ideally suited to higher manufacturing quality levels and higher yields.

SMT production thrives on perfection and dwindles on variations. SMT processes and materials are more sensitive to variation than is IMT. However, the reliability potential for SMT is much higher than that for IMT. To gain all the potential reliability SMT promises requires attention to process and maintaining process discipline. Uniformity and predictability are the chief mandates throughout SMT production.

6.1.4 Impact on traditional quality assurance

The chief responsibility of the traditional quality department, known as *quality assurance* (QA), is rapidly changing. QA could be as busy, and perhaps even more so, during the startup stages of a production project, as they now are at the completion of the project when hardware is being finished.

6.2 SMT Inspection

In the future when perfection and organizational excellence have been fully achieved, QA inspection should no longer be necessary. Between then and now, however, inspection will continue to be needed, but, as the systematic application of statistical process control (SPC) isolates, controls, and ultimately eliminates the root cause of process variations, inspection should diminish as improvements are progressively made over time. In the mean time, data generated by inspection can serve to oversee processes as well as products.

Key to the QA function in SMT is to verify, prior to assembly startup, that the matériel, machines, software, processes, and operator that go onto the SMT assembly line are certified. QA should also monitor the SPC control data during the production operation to verify the proper operation of machines and processes.

6.2.1 Solderability

Solderability is the ability of a surface to be wetted and accept the free flow of molten solder. Solderability of component leads and matching PWB pads is critical to the SMT manufacturing process. Enough emphasis cannot be placed on just how much impact solderability has on SMT manufacturing yields and subsequent reliability.

Solderability defined.   Solderability of component leads and PWB pads is determined by the ability of these metal surfaces to be wetted by molten solder. *Wetting* is the term applied when molten solder freely flows over the surface of heated metal in a relatively uniform, smooth, unbroken coat of solder that adheres to the surface, and forms a feather-blend shape at the edges of the solder extremities. *Dewetting* is a condition that occurs when molten solder has coated the surface and then recedes, leaving uneven thicknesses of solder with obvious areas of very thin coating. *Nonwetting* occurs when large areas of a surface, after having been coated with molten solder, will not adhere to the surface, leaving bare metal exposed (see Fig. 6.1).

Surfaces with unacceptable wetting characteristics can also be detected by examining the edges of the solder that is present. When solder surface tension forces are strong enough to overcome the propensity of solder to flow freely along hot conductors, the surface is most likely covered with a contaminating oxidation or intermetallic layer. Accept/reject criteria, in this instance, are measured by the wetting contact angle (see Fig. 6.2). A special application of wetting angle, a leadless solder joint contact angle (see Fig. 6.3), is the outer edge of the LLCC solder fillet.

Surface contamination and oxidation, of otherwise cleaned and prepared surfaces, often occurs during handling and storage. Storage of solderable components and PWBs should be in an ESD (electrostatic discharge)-approved, temperature- and humidity-controlled area for

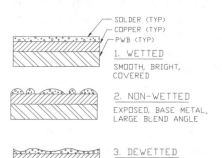

SOLDER (TYP)
COPPER (TYP)
PWB (TYP)

1. WETTED
SMOOTH, BRIGHT, COVERED

2. NON-WETTED
EXPOSED, BASE METAL, LARGE BLEND ANGLE

3. DEWETTED
PULL AWAY, THIN LAYER AREAS

Figure 6.1 Degrees of solderability.

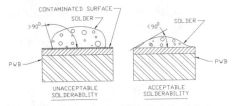

Figure 6.2 Wetting contact angle.

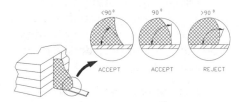

Figure 6.3 Leadless solder joint contact angle.

short-term storage (6 months) and under dry nitrogen for longer periods. Storage should also be in non-paper-based containers to avoid possible exposure to sulfur.

When components and PWBs that have passed solderability tests are removed at one time from longer periods of storage, they should be retested for solderability on a sampling basis.

Protecting PWBs for solderability.   Several methods are used to preserve terminal surfaces for solderability.

1. Applying hot tin-lead dip coating is a good, nonporous protection. Experience has shown that this is the best, long-term protective layer. This coating, however, has a major drawback when applied to FP devices; coplanarity requirements cannot be met by this dip coat technique.

2. Hot-air leveling (HAL) is used, with method 1 above, to overcome the coplanarity deficit. HAL seems to work well for short-term storage periods or on storage, such as with just-in-time (JIT) delivery, but can be ineffective over the long term.

3. Tin-lead plating is easier to apply, costs less, meets coplanarity requirements for FP applications, and is faster than other methods, and consequently is favored by the suppliers. This technique, however, can be unacceptable for very long-term storage (1 year or more). Tin-lead plating, unlike hot-dip coatings, can be plated onto contaminated surfaces, masking a contamination problem that will later emerge at assembly. Plated coatings can also be inconsistent, mixed with additives and brighteners, and fail to protect the base metal surface. The cost-ef-

fectiveness of this protection technique can be positive for the supplier but very negative for the user.

4. Organic coatings over bare copper is another method used to meet coplanarity requirements. This technique is, perhaps, the least effective as a protective coating.

5. Passive chip components, especially those in which terminals are tin-lead-plated can be immersed in hot oil that causes the plated tin-lead to fuse, closing the porous chimneys inherent with plating. This fusing technique has the added advantage, in addition to gaining longer solderability shelf life, of unmasking contaminated surfaces that may be beneath the tin-lead plating as evidenced by dewetting or nonwetting. Hot-oil fusing does, on the other hand, provide the opportunity for the detrimental growth of intermetallics.

The edges of PWB termination pad surfaces could become oxidized and then, consequently, unsolderable if they are not coated. Nonwetting along the edges of terminal pads creates an inherent weakness in the solder joint and can result in failures during thermal cycling.

Nickel barrier coat. When fired metallurgy terminals on components are coated, it is necessary to add a nickel barrier overcoat prior to applying the final protective coat described above. This nickel barrier prevents the migration of noble metals from leaching into the tin-lead coat and contaminating it with undesirable layers of intermetallic. Figure 6.4 shows the growth rate of intermetallics at elevated temperatures, and Fig. 6.5 shows the effectiveness of a nickel barrier coat.

Coating thicknesses. Hot solder dip coats have a long-term effectiveness when the coat is at least 2.50 μm thick and the edges at least 0.38 μm thick. Barrier coats of electroless or electrolytic nickel plate should be at least 1250 μm thick and should cover edges and corners as well.

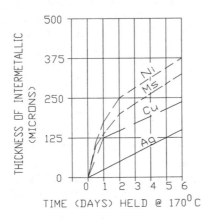

Figure 6.4 Growth of intermetallic.

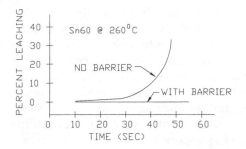

Figure 6.5 Leaching time of Pd-Ag with and without barriers.

When the nickel barrier is too thin, it could be leached and form an intermetallic coformation with the tin and result in unsolderable joints. Nickel over a copper base metalization also prevents the development of brittle intermetallic between the tin and the copper. Nickel can also be used to enhance the solderability of other, less solderable, metals.

Gold plating. When burn-in tests are to be conducted by the user and the burn-in temperature will exceed 150°C, components need to be purchased with gold plate over the nickel barrier coat in lieu of hot solder dip. Gold thicknesses below 10.00 μm can be porous and therefore not be equal to hot solder dip in long-term solderability protection. Gold plating is normally applied in thicknesses ranging from 1.25 to 2.50 μm and, consequently, must be environmentally protected during long-term storage.

Handling and storage. Supplier storage facilities and length of time stored after solderability testing should be reviewed and understood, especially when supplied by a distributor. Storage conditions should be ideally controlled to a relative humidity of 50% and a temperature of 25 ± 5°C. Transportation and storage materials that contain sulfur, silicones, greases, or oils should be eliminated from contact with components and PWBs. Component packages should be sealed and marked. Exposure to the atmosphere, and its inherent moisture, will promote degradation of solderable surfaces. Use of nitrogen-flooded dry boxes is an excellent way to inhibit degradation due to moisture and oxygen.

Accelerated solderability aging. Accelerated age testing is a method used to artificially increase the rate of oxidation on component leads and PWB pads. It is used during incoming inspection to measure the solderability of component leads and PWB pads. Steam aging has been used for this purpose with mixed acceptance. Correlation between length of steam aging and actual storage time, and from batch to batch and facility to facility, has been questionable. Steam aging time, between 8 and 16 h, has been the subject of arguments within the industry.

Solderability testing. Solderability testing has traditionally depended on subjective evaluation of the "dip and look" test method. This method simply consists of dipping the item under test into molten solder and visually examining the surfaces. Very good solderable surfaces and very poor surfaces were easily determined. The problem arises when surfaces which are marginal, especially when the accept/reject decision is viewed by the supplier on one side and the user on the other side. What has been needed, especially for the methodical, predictable, repeatable SMT industry, is an objective test that depends on measured criteria.

Solderability of PWBs. Solderability testing of the PWB can be conducted with the dip-and-look technique in accordance with MIL-STD-202, method 208; MIL-STD-883, method 2003; or the wetting balance test as described in IPC-S-804.

All PWBs, or coupons, need to be baked prior to testing to ensure that moisture, which drastically affects results, has been driven out.

Buoyancy must be compensated when using the wetting balance. Results of the wetting balance test are based on the wetting time and maximum wetting force; therefore, size of the test article, ratio of metal surface to dielectric surface, heat-sink effects, thickness of test specimen, and surface finish of edges all have an effect on results.

The dip-and-look test should not be performed manually but rather, if possible, by a mechanism that inserts the test specimen into molten solder at a constant rate of immersion and withdraws it with accurate measurement of dwell time and insertion angle. Examination should be performed with the unaided eye, or at a maximum of $10\times$ magnification, to verify coverage and consistency.

Plated through-holes should be tested using the timed solder rise test or rotary dip test to determine the wetting and rising through the hole. Variation in thermal mass and the inner-layer configuration often affect the wetting time of the hole.

Solderability of components. Solderability of components can be tested in accordance with MIL-STD-202, method 208; or MIL-STD-883, method 2003 for the dip-and-look and IPC-S-805A for the wetting balance method. The dip-and-look method consists of dipping the component terminal in molten solder and withdrawing it at a constant rate after a set dwell time and examining the surfaces. The wetting balance method suggested by the IPC utilizes a 45° dip, calculates the coefficient of wetting, and then compares it to acceptance levels previously defined.

Until the wetting balance quantitative test method can be verified as being repeatable, the dip-and-look method will continue to be dominant. Suppliers and users should coordinate their solderability test methods to avoid disputes.

Flux used for solderability testing, regardless of the method, has to be controlled to maintain consistency between testers. Consistency of the flux coating and the height of the flux on the test specimen surface needs to be controlled. By using the least active flux, the worst-case condition can be reflected.

Pretinning. Component terminals and PWB pads need to be thoroughly cleaned before they can be properly pretinned. Cleaning techniques should not use mechanical, contact methods but use instead vapor, ultrasonic, spray, or immersion techniques. Immediately after cleaning, the terminals and pads should be hot-solder-dipped. Gold-plated terminals and pads require two dips: the first to be dipped and immediately whipped clean to remove the gold and the second to provide the protective coating. Pretinning can extend solderability shelf life under proper storage conditions for 2 years. All the normal precautions to prevent thermal shock, such as preheating and natural cooling, must be adhered to when pretinning.

Rework of solderability rejected parts. Components and PWB can be reworked by cleaning, fluxing, and resoldering, but all guidelines and normal soldering standards must be maintained.

6.2.2 Inspection of SMT components and PWBs

Inspection of SMT components and PWBs is not unlike inspection of IMT components and PWBs except for size and packaging. Determining component and PWB pedigree, such as screen tests, burn-in, solderability tests, fabrication lot and date, storage and transportation history, supplier, and name and address, and general workmanship examination are all similar to those for IMT. Because of the size of parts, visual examination may require the use of 3 to 10× magnification. SMT uses unique component shipping and handling packaging (see Chap. 3). Since these packages are designed for direct placement into the assembly line, it is important to determine the condition of each prior to submitting it for assembly-line usage. In tape-and-reel systems, the cover seal tape should be in place with components in the desired orientation. Cover tape peel force should be verified. When paper tapes are used, determine whether the paper has a negative effect on the solderability of the leads and that loose paper flake will not cause a clog problem for assembly-line machinery. Determine that waffle trays have not been dented and that all materials for sensitive components are properly rated for ESD protection.

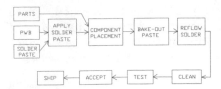

Figure 6.6 Basic SMT assembly sequence.

6.2.3 Solder paste inspection

Solder paste (sometimes referred to as a "cream") has an unusually short shelf life, compared to those of other solders, and is very sensitive to moisture, oxygen, and gravity. It is a thixotropic paste, which should be uniform in consistency and free from contaminants or debris. Slight settling or separation of solder particles should not be cause for rejection, provided uniformity can be reestablished by gentle stirring. See IPC-SP-819 for general requirements and inspection of solder paste. Solder paste tends to be batch-sensitive and therefore, until proved otherwise, each lot should be inspected and tested for metal content by weight, viscosity, spread and/or slump, solder ball, tack, and wetting. Contents of each container should be visually examined for consistency and general appearance as a minimum.

Ideally a sampling of the paste should be processed through the actual screen, stencil, or syringe dispenser and then reflow-soldered in the production machine before acceptance.

6.2.4 Assembly process inspection

Solder paste application, component placement, soldering, and cleaning are all critical aspects of the SMT assembly fabrication process (see Fig. 6.6).

Paste application. Verify the paste to pad alignment and solder paste volume to be correct and accurate. Verify the adequacy of paste viscosity, seeing that the paste did not slump, yet totally covered its intended pad. Check for tackiness after bake-out (if a separate bake-out step is included). Examine the screen and/or stencil for residue paste.

Component placement. Besides general alignment, polarity, and function pin orientation, the critical features are coplanarity with no gap between the leads and PWB pads with thicknesses greater than 0.10 mm. Verify the attitude and level of passive chips.

Soldering. Inspection for solder defects includes electrical and mechanical defects and workmanship. Electrical defects and causes consist of shorts and opens (short circuits and open circuits) as follows:

1. Shorts (bridges)
   a. Excessive volume of solder
   b. Misplaced components
   c. Designed-in high-density conductors and/or pads
   d. Poor paste screening alignment
   e. Solder paste slump
   f. Inadequate solder reflow profile
2. Shorts (solder balls)
   a. Failure to bake out all paste solvents
   b. Poor paste screening alignment
   c. Improper rheology of paste organics
   d. Oxidation of paste solder particles
3. Opens
   a. Thermal shock
   b. Poor solderability of terminals and/or pads
   c. Poor solder reflow temperature profile
   d. Incorrect reflow machines, shields, and/or adapters
   e. Insufficient solder volume
   f. Uneven solder volumes for component
4. Potential field failures
   a. Partially attached solder balls—loose in vibration
   b. Flux residue—electromigration and corrosion
   c. Inadequate solder joints—short fatigue life
   d. Poor thermal design management—thermally cycled cracked solder joints
   e. Solid-state diffusion growth of intermetallics, progressive development of weaker solder joints
5. Workmanship
   a. Dull solder joints—solder oxidation, slow cool-down
   b. Excessive skew of components
   c. Local PWB discoloration
   d. Nonflux stains
   e. Solder voids in outer surface of solder joint

Assembly inspection. The most critical solder joints in SMT are those that attach LLCC devices. See Fig. 6.7 for accept/reject profile of the LLCC solder joint configuration.

Two important determinations can be made from the resultant data of postsoldering assembly inspection. First, individual product acceptance is decided. Second, the more important determination, is a measure of the machines and processes for the short-term purpose of verifying the current production operations and the long-term purpose of seeking ways to improve throughput time and yields.

The following criteria present variables in the attachment of compo-

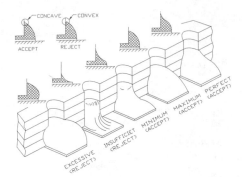

Figure 6.7 LLCC solder joint accept/reject criteria.

nents to SMT assemblies. These criteria are based on IPC's standard number NTL-STD-SOLD (proposed).

LLCC components. The most critical solder joints in SMT are those used to attach LLCC active component devices to PWBs. Figure 6.7 presents the accept/reject criteria for judging LLCC solder joints based on the joint's configuration profile.

LDCC components. For LDCC active component fabrication of assemblies, see Fig. 6.8 for J-leaded devices and see Fig. 6.9 for gull-wing-leaded devices. See Fig. 6.10 for butt (I)-leaded devices.

Discrete passive chip components. Acceptable variations in discrete passive chip component attachment are shown in Fig. 6.11. Figure 6.12 shows the frequently encountered rejectable tombstone attitude of chip components and skewed chip components.

Postsolder cleaning.   Postsolder cleaning SMT assemblies, in general, is far more difficult than cleaning IMT assemblies. With SMT, component standoff height is far less, component density is more, large square components are used, component lead pitch is less than half, leadless active devices are present, and double-sided assemblies are used. All these features tend to reduce the effectiveness of established cleaning techniques.

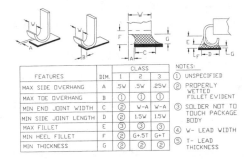

FEATURES	DIM.	CLASS 1	CLASS 2	CLASS 3
MAX SIDE OVERHANG	A	.5W	.5W	.25W
MAX TOE OVERHANG	B	①	①	①
MIN END JOINT WIDTH	C	②	W–A	W–A
MIN SIDE JOINT LENGTH	D	②	1.5W	1.5W
MAX FILLET	E	③	③	③
MIN HEEL FILLET	F	②	G+.5T	G+T
MIN THICKNESS	G	②	②	②

NOTES:
① UNSPECIFIED
② PROPERLY WETTED FILLET EVIDENT
③ SOLDER NOT TO TOUCH PACKAGE BODY
④ W– LEAD WIDTH
⑤ T– LEAD THICKNESS

Figure 6.8 J-lead joint specifications.

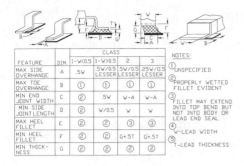

FEATURE	DIM	CLASS				NOTES:
		1-W<0.5	1-W>0.5	2	3	① UNSPECIFIED
MAX SIDE OVERHANG	A	.5W	.5W/0.5 LESSER	.5W/0.5 LESSER	.25W/0.5 LESSER	② PROPERLY WETTED FILLET EVIDENT
MAX TOE OVERHANGE	B	①	①	①	①	③ FILLET MAY EXTEND INTO TOP BEND BUT NOT INTO BODY OR LEAD END SEAL
MIN END JOINT WIDTH	C	②	.5W	W-A	W-A	
MIN SIDE JOINT LENGTH	D	②	W/0.5	W	W	
MAX HEEL FILLET	E	②	②	③	③	④ W-LEAD WIDTH
MIN HEEL FILLET	F	②	②	G+.5T	G+.5T	⑤ T-LEAD THICKNESS
MIN THICKNESS	G	②	②	②	②	

Figure 6.9 Flat ribbon and gull-wing lead joint specifications.

W = LEAD WIDTH
T = LEAD THICKNESS

FEATURE	DIM.	CLASS 1	CLASS 2	CLASS 3
MAX. SIDE OVERHANG	A	1/4 W	ZERO	ZERO
MAX. TOE OVERHANG	B	ZERO	ZERO	ZERO
MIN. END JOINT WIDTH	C	3/4 W	W	W
MIN. SIDE JOINT LENGTH	D	2 T	3 T	4 T
MAX. FILLET	E	NOTE 2	NOTE 1	NOTE 1
MIN. FILLET (mm)	F	0.5	0.5	0.5
MIN. THICKNESS	G	NOTE 2	NOTE 2	NOTE 2

NOTES. 1. MAX. FILLET OK INTO RADIUS, NOT BODY.
2. PROPERLY WETTED FILLET EVIDENT.

Figure 6.10 Butt-lead joint specification.

FEATURE	DIM.	CLASS			NOTES:
		1	2	3	① 1/2W, 1/2P, 1.5 WHICHEVER IS LESS
MAX SIDE OVERHANG	A	①	①	②	② 1/4W, 1/4P, WHICHEVER IS LESS
MAX TOE OVERHANG	B	TANG. TO LAND	TANG. TO LAND	TANG. TO LAND	③ 1/2W, 1/2P, WHICHEVER IS LESS
MIN END JOINT WIDTH	C	③	③	④	④ 3/4W, 3/4P, WHICHEVER IS LESS
MIN SIDE JOINT LENGTH	D	⑤	½T	½T	⑤ WETTED FILLET EVIDENT
MAX FILLET	E	⑥	⑥	⑥	⑥ MAX FILLET, EXCEPT SOLDER NOT TO ENCASE TERMINAL
MIN FILLET HEIGHT	F	⑤	⑦	⑦	⑦ G + 1/4H, 0.5 WHICHEVER IS LESS
MIN THICKNESS	G	⑤	⑤	⑤	

Figure 6.11 Discrete chip soldering standard.

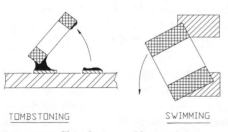

TOMBSTONING   SWIMMING

Figure 6.12 Chip devices soldering defects.

Low component standoff height, although beneficial for thermal management and volume density, is detrimental to thorough cleaning beneath the components. This same low standoff feature also makes it more difficult for contaminants to get beneath the components. Nevertheless, contaminants do get beneath components, and cleaning is necessary.

Getting cleaning fluids under the components can be accomplished somewhat with the use of wetting agents and high-pressure jets. Removal of the contaminants and cleaning fluids is the more difficult part of the problem. Contaminants are best removed by flooding them away. Tight lead spacing all around the component body acts as a fence, inhibiting the flow from backwash. IMT components are positioned with a standoff height of 0.38 mm or more above the PWB surface. Standoff height of SMT components is often less than 0.25 mm.

Manufacturing residues and contaminants, on postsoldered SMT assemblies, consist of elements introduced by the solder operations and others that may be there as a result of earlier processing and handling. Some of the contaminants include solder flux residue, potentially the most damaging in the long run; surfactants; neutralizers; earlier processing salts; and oils, grease, wax, metal oxides, and lubricants. Ionized polar soils (plating salts left over from earlier fabrication operations, fingerprint salts, and flux activators are some of the common soils), are substances that can align themselves in an electrical field and carry electric currents, which can be deleterious to finished assemblies. These ionized elements are quite reactive and will react with PWB metals to form corrosion products. These same residues can also become conductive in the presence of moisture and create a high-resistance leakage path across the circuit.

Nonpolar compounds, such as oils, greases, rosin, and other waxes, can become insulation films on contact surfaces on devices such as connectors, switches, and potentiometers.

Cleaning solvents. Cleaning solvents are either polar or nonpolar. Polar solvents, such as alcohols or ketones, are used to remove polar residues. Nonpolar solvents, chlorinated and fluorinated solvents, are used to remove nonpolar soils. Electronic residues are generally composed of a combination of polar and nonpolar parts. Cleaning solvents, therefore, need to be a blend of polar solvents and nonpolar solvents to be truly effective.

Azeotropic solvent systems, which have exactly the same compositions in liquid and vapor forms at the boiling point, are used in vapor degreaser cleaners. This azeotropic feature allows the solvents to be continuously cleaned as the distillation process, produced by the vapor-condensation cycle, functions as a surrogate filter.

Nonpolar solvents can be subclassified as chlorinated and fluorinated solvents. Chlorinated solvents—perchloroethylene, methylene chloride, and 1,1,1-trichloroethylene—are more aggressive, are more toxic, and have lower evaporation rates and higher boiling temperatures than do fluorinated solvents. These solvents are normally used in cold cleaning applications. Fluorinated solvents are milder, have lower solvency power, have higher evaporation rates and lower boiling points, and generally are less toxic. These solvents are used in vapor cleaning operations.

Aqueous cleaning. Rosin fluxes, not normally soluble in water, can become soluble by adding saponifiers to the water. The reaction between the rosin flux, the water, and the saponifier produces a rosin soap which is readily soluble in water.

Caution. This cleaner, itself, is highly ionic and if trapped beneath SMT components can cause a serious contamination problem. Copious amounts of rinse water are needed to ensure adequate cleaning.

Cleaning methods

1. *Cold cleaning.* Cold cleaning (room temperature to 100°C) is usually done in an in-line, conveyorized machine using chlorinated solvents. This technique is used for high-volume, lower-reliability production. Wave cleaning methods that use spray, rotating brushes, and immersion dipping are variations of cold cleaning.

2. *Vapor cleaning.* This cleaning system uses fluorinated solvents and is very effective for SMT high-density assemblies.

3. *Ultrasonic cleaning.* This is a very effective cleaning system within certain bounds.

Alleged ozone depletion. Because of the alleged reduction in the ozone layer in the upper atmosphere due to CFCs, allegedly attributed to the inadvertent release of chlorinated and fluorinated solvents (CFCs) into the lower atmosphere by the electronics industry and others, prime cleaning fluids are, following the Montreal Protocol, scheduled for phase-out in the immediate future. The electronic industry has been searching for alternative cleaning fluids, alternative processes, or the elimination of the need to clean. The goal of the industry has been to find and adapt a process that would produce a product which is as reliable as those cleaned with CFC solvents, without incurring large cost impacts.

Three prominent alternatives to CFC solvents are currently in development or use: (1) alternate solvents, (2) no-clean fluxes, and

(3) aqueous cleaning. These and other cleaning agents and methods are described in the following paragraphs.

1. *Alternate solvents.* There are no commercially available alternative solvents that are completely safe for the environment and for people. Chlorinated solvents are as good as CFCs but unstable in use and have questionable toxicity. Alcohol is an adequate cleaner but is flammable. Materials are being developed, but none that appear to be equivalent. Gaining final approval, in this politically charged social environment, could prove to be the biggest challenge facing alternative solvents.

2. *Terpenes.* Terpene is a hydrocarbon-based solvent, readily available from turpentine and similar to rosin flux in chemical composition. It contains no carbon-chlorine bonds, is reported to clean as well as CFCs, but has mixed reviews on its production readiness status. It has an excellent affinity for quickly and efficiently dissolving rosin fluxes, and continues to clean even when heavily contaminated. Terpene removes both polar and nonpolar contaminants. It has a low viscosity, which permits it to flow beneath SMT components, and is noncorrosive. Terpene is environmentally friendly, has a neutral pH, and contains no chlorine. Production drawbacks include a very low flashpoint (<37°C), which presents a safety problem on the assembly line, tendency to foam if sprayed when diluted, a pungent odor, and low viscosity, thus making ultrasonic application unlikely. Since water is the rinsing agent, and water is unlikely to move out easily from under SMT components, the surfactants from the terpene suspended in the rinse water could be left on the assembly beneath SMT components. Terpenes tend to foam with water intermixing and, finally, material compatibility has not been confirmed.

3. *No-clean flux.* This CFC replacement alternative uses mild flux that leaves a benign residue and therefore does not entail cleaning. However, with this approach the flux activity is less than that normally needed for ordinary assemblies. To make no-clean flux work, extraordinary process steps are needed prior to and during the soldering operation to rid the PWB pads and component leads of oxides and other contaminants and thereby to reduce the need for active flux. PWB pads and component leads should be certified for solderability just prior to the soldering operation; and, if wave-soldered, the soldering needs to be done in an oxygen-free nitrogen atmosphere. Even with these precautions doubt persists regarding the adequacy of using an unusually mild flux and the long-term effect of leaving flux residue on the assemblies, no matter how benign the flux is to begin with. Flux residue can interfere with subsequent fabrication processes, such as in-circuit testing or conformal coating, and if cleaning is eliminated altogether, then other contaminants and debris, such as solder balls, fibers resulting from

PWB router process, oils, and fingerprints will not be removed and could become nearly as serious a problem as flux residue.

4. *Organic water-soluble flux.* Water-soluble flux is an older compound that has been revised recently in response to the alleged CFC ozone depletion problem. With this technique flux residues are removed by water in lieu of chemical solvents. Organic water-soluble fluxes are very active, higher-powered fluxes with complex mixtures of organic acids, halide salts of organic acids and amines, wetting agents, and solvents. Residue from this type of flux has an inordinate propensity to absorb water, greatly enlarging its volume and increasing its mobility.

Ideally, water-soluble flux should be used for assemblies that do not have components, or areas, that can entrap contaminated cleaning or rinse water, such as open relays, transformers, potentiometers, porous components, or stranded wiring. High-density SMT assemblies could be a problem. Because of the amount of salts within the flux, insulation resistance of porous paper-based and phenol-based boards can be detrimental.

5. *Aqueous and semiaqueous cleaning.* Aqueous cleaning uses ordinary water and is used primarily to remove water-soluble flux. Water is a very effective solvent, but with postsolder cleaning, copious amounts of water are needed. With the addition of a detergent or saponifier in the cleaning cycle, nonionic residue (rosins), as well as the ionic residues, can be removed. Detergents also reduce water surface tension, enhancing penetration beneath SMT components with low standoff height. Wetting agents are added to the rinse cycle to help water to rinse beneath the low components. Except for strong capillary pressure, the high surface tension of water would prevent its penetration between the PWB and low standoff components.

Measuring cleanliness. To maintain control over the process and to ensure satisfactorily cleaned assemblies, a postcleaning inspection is performed. Inspection has been performed as follows:

1. Visual examination is performed to detect residue deposits.
2. Ultraviolet light is used to measure rosin residue through fluorescence.
3. Acetonitrile is employed to detect nonionic organic residues.
4. Infrared spectrum of the extract is used to identify specific contaminants.
5. Water conductivity tests first measure clean deionized water, then after extraction of water-soluble ions from the surface of the assembly, compare the two measurements; the differences indicate the measure of cleanliness.

6. Ionographs are also used to measure the conductivity of ionic soils in the rinse solution extracted from the cleaned assembly surface.

7. Surface insulation resistance (SIR) measurement is another method used. To evaluate cleaning beneath large SMT components, components are removed after cleaning and SIR measurements made on the exposed surfaces.

6.3   Automatic Inspection

Manual inspection of SMT assemblies can quickly become boring and fatiguing. The sheer quantity of small, densely populated components can exact an overwhelming toll on an inspector, severely impairing judgment. Inspectors can visually inspect 20 to 30 joints per minute. With a typical SMT assembly of 4000 solder joints, inspection of solder joints alone can average 2.5 h per assembly.

Depending on inspection rates, the SMT inspector, to remain effective, requires some level of help from machines designed specifically for quality inspection. With the machines now available, the human (operator)/machine ratio can span the full range from 1/0 to 0/1.

Quality-oriented machines are available as visual only, sensory only, electromagnetic ray, and computer-supported. Machines with one or a combination of two or more of these features can be obtained. Some machines can be tied to computer networks that interconnect to one or more company plants in separate locations. CAD (computer-assisted design) data from the design process can be automatically fed into self-programming machines, and automatic output inspection data can be tied to process control stations and quality record-keeping locus.

The overwhelming majority of process defects reported by industry fit the following categories:

1. Solderability—whole or partial leads and pads wetted, dewetted, or nonwetted

2. Coplanarity—individual component leads and PWB pads on different planes

3. Solder paste quality, application, and quantity

4. Reflow profile of final solder joint

5. Component placement accuracy

The last four categories are ideal for automatic inspection using the newer multisensor, computer-backed, quality machines. These newer machines have a high throughput rate with a defect-found rate that measures better than 95 percent and a very low false-reject rate.

Figure 6.13 In-line, real-time, fully automated, conveyorized, solder joint radiographic (X-ray) inspection station. (*Courtesy of IRT Corp.*)

Visual. Visual aid machines can be as simple as an apparatus with a magnifying glass or as complex as a machine that combines optical scanners, image enhancement, and closed-circuit televison.

Sensory. Sensory aid, in the form of light and sound reflections, thermal monitoring with infrared and lasers and ultrasonic devices, are all moving the human/machine ratio toward 0/1.

Electromagnetic rays. X-Ray machines are indispensable as inspection devices during assembly process development phases. X-Ray machines can be valuable for high-rate production (see Fig. 6.13).

6.4 Production Certification

Uniformity and predictability are the major production process goals in SMT. As these goals are met by continual improvements in the manufacturing processes and as similar goals, imposed on matériel, also experience improvements, the main focus of QA can shift from 100 percent postproduction inspection of the end product to 100 percent preproduction certification of the machines, processes, and matériel that constitute the end product.

X-Ray machines serve to gain invaluable insight in soldering processes, as the process is being developed and certified (see Fig. 6.14).

Kaizen. Key to making the shift in emphasis from postproduction inspection to preproduction is based on the adoption of Japan's Kaizen philosophy. Kaizen strategy is the single most important quality concept in the Japanese industry: it is key to Japanese competitive success. By implementing Kaizen, Japanese industry was able to transcend market changes that went from abundant low-cost resources, quantity-oriented customers, and a rapidly expanding market

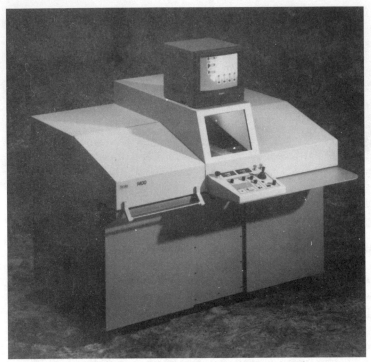

Figure 6.14 Real-time X-ray imaging system. (*Courtesy of Nicolet Test Instruments.*)

to a market of higher-cost resources, overcapacity of production facilities, increased competition in saturated and dwindling markets, sharply changed consumer values for higher quality, and a need to introduce new products more rapidly at a lower break-even point.

Kaizen is a customer-driven strategy for improvement. In contradiction to the traditional philosophy that says "if it ain't broke, don't fix it," Kaizen philosophy says "gradual, unending improvements, even in the little things, lead to ever-higher standards and customer satisfaction." The traditional philosophy is based on the belief, and experience, that changes cause more problems than they fix. Kaizen does not deny the problems that change brings; it says that with the correct attitude and total staff involvement and attention, industry can work past problems and produce positive quality improvements and, at the same time, reduce throughput time and reduce costs. The Japanese and a number of American companies have proved this to be the case with higher profits, higher sales, and less rework staff.

Traditional production philosophy holds cross-functional production problems to be seen in terms of conflict resolution, where those responsible for the problem are punished. In Kaizen, cross-functional problems are resolved on a systematic and collaborative approach, with no one being punished and those responsible for the solution being recognized in a positive way. In Kaizen, the pursuit of higher quality is a never-ending process.

Malcolm Baldrige National Quality Award.   The U.S. Congress created the Malcolm Baldrige National Quality Award (MBNQA) in 1988 to energize the U.S. industry into pursuing higher quality standards. Administrated by the Commerce Department, the award has quickly become, for those large and small companies who choose to pursue it, at once a map and motivator to upgrading the company's ability to produce much higher-quality products and/or services. Winning the MBNQA results in immediate worldwide product recognition and sales as well as other industrial rewards. But winning is not prerequisite to achieving industrial rewards. The pursuit alone results in quality improvement and higher customer acceptance.

The MBNQA measures a company's performance in seven major categories:

1. *Leadership.* Senior management's success in establishing and sustaining a companywide acceptance of a quality-conscious culture.

2. *Information and analysis.* Measurement of the company's SPC, and other data collection systems, for quality improvement and planning.

3. *Planning.* Measurement of the integration of quality requirements into company business plans.

4. *Human resources utilization.* Effectiveness of efforts to involve total staff in quality.

5. *Quality assurance.* Measurement of company's quality system in assuring quality control of all operations.

6. *Quality assurance results.* Quantitative improvements achieved by the quality system.

7. *Customer satisfaction.* Measurement of the effectiveness of the company's ability to determine and meet customer requirements.

There are three categories and up to two awards given in each category, if there is a clear winner. Listed below are the three categories:

1. Large manufacturing company and subsidiaries

2. Large service company and subsidiaries

3. Small manufacturing and service company (less than 500 employees)

Each point of the seven (7) criteria can receive a score of up to 1000. Judgment is biased toward measurable results, not just systems and programs.

The board of examiners consists of 140 quality experts from industry, universities, professional organizations, and trade organizations. Written summaries of strengths and weaknesses found by the examiners in the company's quality program are made available to the participating company. Winners are expected to share information regarding their quality strategies with other U.S. companies. Applicants are measured against worldwide quality levels.

Note: To apply, contact Malcolm Baldrige National Quality Award Chairman, National Institute of Standards and Technology, Gaithersburg, Maryland 20899.

Bibliography

Austen, Paul, and Scott Downey: "Don't Forget Safety," *Surface Mount Technology,* June 1992, pp. 32–36.

Borek, Keirn: "A Clean Break from CFC's," *Circuits Assembly,* February 1992, pp. 43–48.

Burns, R. Allen: "A Machine Vision Solution to SMD Inspection," *Surface Mount Technology,* November 1991, pp. 44–46.

Capillo, Carmen: *Surface Mount Technology,* McGraw-Hill, New York, 1990, pp. 112–146.

Fauber, William J.: "In Pursuit of ISO-9000," *PC FAB,* June 1992, pp. 52–54.

Haavind, Robert: "New Thinking on Quality," *Electronic Business,* October 16, 1989, pp. 24–28.

———, "Baldrige Quality Award: Industry's Version of Oscar," *Electronic Business,* October 16, 1989, pp. 34–36.

Hayes, Michael E.: "Cleaning SMT Assemblies without Halogenated Solvents," *Surface Mount Technology,* December 1988, pp. 37–40.

Heffner, Kenneth H., Jack C. Brand, Barbara Grosso, and Richard Terrell: "The Effect of Flux Residues on Long and Short Term PWB Insulation Resistance Assays," *Technical Review* (IPC), December 1988, pp. 14–27.

Hodson, Timothy L.: "Aqueous System Cleans Us at Cal Comp Display Products Group," *Electronic Packaging and Production,* December 1991, p. 47.

Hollomon, James K., Jr.: *Surface Mount Technology for P.C. Board Design,* Howard W. Sams, Indianapolis, Ind., 1989, pp. 212–264.

Kumar, Vijay: "Total Quality Management for SMT," *Circuit Assembly,* December 1991, pp. 54–56.

Lucas, Alfred R.: "Motorola's Kaizen Brigade," *Design News,* February 10, 1992, pp. 98–104.

Mindel, M. J.: "Flux Corrosion Test Reflects Accurate Field Operating Conditions," *Electronic Packaging and Production,* December 1989, pp. 76–77.

Munson, Terry: "Cleanliness Testing for the 90's by Ion Chromatography," *Technical Review* (IPC), August 1991, pp. 19–26.

———, "Cleanliness Testing for the '90s," *Circuits Assembly,* August 1991, pp. 24–28.

Paul, John T.: "Ultrasonic Cleaning—Can Current Technology Meet the SMT Challenge?" *Technical Review* (IPC), October 1989, pp. 15–23.

Richards, B. P., P. Burton, and P. K. Footner: "Does Ultrasonic Cleaning of PCBs Cause Component Problems: An Appraisal," *Technical Review* (IPC), June 1990, pp. 15–27.

———, "Damage-Free Ultrasonics Cleaning Using CFCs, Aqueous, Semi-Aqueous Solvents," *Technical Review* (IPC), March 1991, pp. 26–30.

Shandle, Jack: "Cutting out Cleaning," *Electronics,* January 1991, pp. 41–44.

Stein, Howard, "Laser-Based 3-D Lead Inspection," *Surface Mount Technology,* June 1992, pp. 39–42.

Stont, Gail: "ISO 9000 Driven by Threats and Promises," *Surface Mount Technology,* June 1992, p. 7.

Swenson, Raymond: "Validation of an Automatic Solder Joint Quality Measurement System," *Surface Mount Technology,* October 1991, pp. 25–36.

Szymanowshi, Richard A.: "Fluxing Options for CFC Elimination," *Technical Review* (IPC), April–May 1989, pp. 19–23.

Troebel, Thomas W.: "High-Volume, High-Speed, Solder Joint Inspection," *Circuit Assembly,* November 1991, pp. 42–47.

Turbini, Dr. Laura J.: "Cleaning Issues," *Circuits Assembly,* August 1991, p. 36.

Wenger, George M., and Gregory C. Munie: "Defluxing Using Terpene Hydrocarbon Solvents," *Technical Review* (IPC), November 1988, pp. 17–23.

Wesselman, Carl: "Cleaning 1991: The Picture Stabilizes," *Surface Mount Technology,* July 1991, pp. 26–30, 53–55.

Whiting, Rick: "Baldrige Application Process: A Long, Rigorous Examination," *Electronic Business,* October 16, 1989, pp. 44–45.

Wrezel, J. A., R. C. Pfahl, Jr., and L. R. Hagner, "Semi-Aqueous Cleaning Replaces CFC Defluxing: A User's Perspective," *Electronic Packaging and Production,* April 1992, pp. 43–48.

Chapter

7

Electrostatic Discharge

7.1 Industry Impact

In the electronics industry electrostatic discharge (ESD) is known as the "hidden killer." It generally operates silently without external manifestation. Each year it is estimated that the cost of ESD damage to electronics is in billions of dollars of direct damage to electronic devices. Electrostatic damage operates at all phases of the fabrication cycle and at all levels of the system from single components to completed end products. The ESD problem has become so acute, especially in the ever-diminishing sizes of SMT components and assemblies, that everyone in the production cycle, from the component supplier to the final assembly inspector, must participate in the solution.

7.2 Semiconductor ESD Damage

7.2.1 Causes

Damage to semiconductor devices can be caused by direct ESD through the device leads when inadvertently contacted by a charged surface, or it can be induced while the device is in the midst of a high electrostatic field. Damage can be characterized by a functional distortion, or total loss, of information or a hard failure rendering the component inoperable.

7.2.2 Damage description

Electrostatic discharge currents may flow directly into vulnerable elements of ESD-sensitive circuits or may enter indirectly through capacitive, magnetic, or electromagnetic coupling. In either event, ESD damage can easily distort or destroy the device. Complementary metal oxide semiconductor (CMOS) devices have submicron gate channels with thin silicon dioxide insulation layers that are easily destroyed by ESD discharges. ESD transient energy can flow through the oxide layer and vaporize a section in one of the metalization circuit trace or bond pads.

7.2.3 Categories of failure

Spark-type ESD discharge events take place in time scales as low as a few nanoseconds and can involve currents as high as a few amperes. The damage may cause an immediately detectable failure or, more often, degrade the device to the point where it functions intermittently or with a latent defect that eventually fails during operational usage.

Latent defects are the worst kinds of damage. Damage of this type is not severe enough to be immediately apparent but weakens the device to the point that continued usage will soon cause an intermittent operation or a catastrophic failure in the field. Latent defects impact warrants and can jeopardize critical flight avionics.

7.3 Damage Sensitivity

All electronic devices are vulnerable to ESD damage to one degree or another, depending on the device's particular microelectronic technology. Each technology family has its own "damage sensitivity" level in terms of breakdown voltage across the thin oxide layers and interconnecting paths. Table 7.1 lists the ESD voltage sensitivity levels of the various semiconductor groups. Gallium arsenide devices are particularly sensitive to ESD because of the thinness of their oxide layers. The range of voltages capable of damaging these devices is only 30 to 60 V.

Some of the ESD-sensitive components include microcircuits, discrete active devices, chip resistors, hybrids, and crystals. DoD

TABLE 7.1 Device Sensitivity to ESD

TYPES OF DEVICE/TECH.	ESD POTENTIAL VOLTAGE RANGE (V)
VMOS	30 TO 1800
EPROM	100 MAX.
MOSFET	100 TO 200
SAW	150 TO 500
CMOS	250 TO 2000
BIPOLAR	380 TO 7000
SCR	680 TO 1000

Standard 1686 and DoD Handbook 263 list the following three levels of component sensitivity:

Class I—0–999 V

Class II—1000–3999 V

Class III—4000–15,000 V

Class I components are the most sensitive and therefore require the greater amount of ESD protection.

7.4   How ESD Is Generated

Static electricity is generated by friction or separation of materials, flowing liquids, vapors or particles in gases. Common sources within the electronics industry that produce an electrostatic potential, beyond the workforce personnel themselves, include plastics in the form of solid objects, clothing, film layers of paints and waxes and any spray-applied finish or rinse. It includes processing equipment such as dryers, heat guns, conveyor belts, wave-soldering machines, component placement machines, handling carts rolled over vinyl flooring, removal or placement of assemblies in plastic film bags, and untaping the cover layer tape from reeled components. The processing equipment listed here, and other equipment, can generate electrostatic fields that need not directly contact an ESD-sensitive component to damage it.

7.5   Influence of Materials on ESD

7.5.1   Material reaction

Different materials react differently when rubbed, generating either a positive or a negative electrostatic charge. Electrons, being negatively charged, are transferred from one surface to another, leaving the surrendering surface positive and making the acquiring surface negative. Electricity produced in this way, known as *triboelectricity,* a term taken from the Greek word *tribos* meaning a rubbing, builds and accumulates on surfaces which do not have conductive discharge paths to neutralizing surfaces. The magnitude of a charge generated in this way is a function of the rubbing pressure, the speed of rubbing, the amount of humidity, and the propensity of the material to surrender or acquire electrons.

7.5.2   Triboelectric Series

Materials have been listed in a series called the *Triboelectric Series* which places them in relative positions to one another according to their propensity to surrender or acquire electrons. At the top of the series is

TABLE 7.2   ESD Triboelectric Series

AIR (+) (POSITIVE)	COTTON (REF. POINT) (+/-)	
HAIR	STEEL	
ASBESTOS	WOOD	
GLASS	AMBER	
MICA	HARD RUBBER	
HUMAN HAIR	NICKLE/COPPER	
NYLON	BRASS/SILVER	
WOOL	GOLD/PLATINUM	
FUR	ACETATE/RAYON	
LEAD	POLYESTER	
SILK	POLYURETHANE	
ALUMINUM	PVC (VINYL)	
PAPER	SILICON	
COTTON (REF. POINT) (+/-)	TEFLON (-) (NEGATIVE)	

the material with the highest potential positive charge, that material which most readily gives up electrons, and at the bottom of the list is that material with the highest potential for negative charge, that material which most readily acquires electrons. The benefits of this series are to gain insight into which materials, when moved in close proximity to one another, would generate the most ESD hazard (see Table 7.2).

Insulating materials acquire a charge when rubbed against other insulating materials or metals, even if the metal is grounded. Total prevention of charging cannot be accomplished by the selection of like materials; materials will acquire charges even when rubbed against themselves.

7.5.3   Neutrality of cotton

Notice that cotton is half way between the two extremes in the Triboelectric Series. Cotton is considered neutral and explains why it has been the cloth of choice for laboratory smocks and personnel clothing. Notice the relative position of nylon to polyester, two popular clothing materials, and how potentially damaging they could be.

Electron flow from an accumulated charge on nonconducting materials will occur in the vicinity of the contact point made on the charged surface and a conducting path. This restrictive flow makes it difficult to neutralize static charges of large areas on nonconducting surfaces.

Newer smocks are made of lightweight washable and durable polyester fabric with continuous conductive threads woven throughout and connected to a wrist-snap terminal at the cuff of one of the sleeves to dissipate residual static charges that accumulate on personnel and their clothing and that cannot otherwise by removed with normal personnel grounding devices.

7.6   Causes of ESD

7.6.1   People—a major source of ESD

The major source of static charge buildup in the working environment is the movement of people. Accumulated charges on personnel and

TABLE 7.3   Effects of Humidity on ESD
Voltage Generation

SOURCE	LOW RELATIVE HUMIDITY 10-20% RH	HIGH RELATIVE HUMIDITY 65-90% RH
WALKING ON CARPET	35 KV	1.5 KV
WALKING ON VINYL	12 KV	0.3 KV
WORKER AT BENCH	6 KV	.1 KV
PLASTIC INSTRUCTIONS	7 KV	.6 KV
POLYETHYLENE BAGS	20 KV	1.2 KV
POLYETHANE FOAM	18 KV	1.5 KV

their clothing in dry climates can reach 35,000 V. This value is more typically in the 1000-V range in regions of higher relative humidity (RH) (see Table 7.3). Sparks generated by static electricity begin to occur at 3000 V. However, nonspark discharge at lower voltages is nearly as damaging and all the more pervasive.

7.6.2   Other major sources of ESD

Other prominent sources of ESD include carpets, floor tile, non-ESD packaging materials, insulated workbench tops and chairs, doorknobs, shoes, CRT screens, and ungrounded automatic production machines and equipment.

Assemblies can accumulate a charge as they are automatically moved from one machine to another. Sometimes even grounded machines are a problem: the mainframe may be grounded but add-on mechanisms may not be appropriately connected to the grounded mainframe with a bonded joint. Airflow in convection ovens and IR solder reflow machines, spraying of conformal coating, and stripping of insulation tape off assemblies can all generate damaging ESD charges.

7.7   ESD Control

ESD control is based on (1) prevention and (2) protection.

7.7.1   Prevention of ESD

ESD is another of those negative industrial issues that requires the conscious concern and attention of upper management. Because it is an insidious problem without bells or whistles, everyday concern, by personnel, for the potential hazards of ESD naturally and quickly subsides, leaving the prevention side of control ineffectual. The entire workforce needs to be informed and continuously reminded about the hazards of ESD. Those individuals with hands-on exposure require a more detailed and practiced training in the proper handling and storage of vulnerable electronics. Company rules, regulations, and standards concerning ESD need to be established and enforced. The potential loss of electronic products, productivity, and company reputations will not allow less than a relentless and arduous pursuit of total commitment by management and the workforce to ESD prevention.

ESD plans.   ESD control plans should be initiated for each company site from incoming receiving through delivery of the end product. Start by evaluating the operations at each site. Determine the needs for equipment, facilities, and training. Formulate plans tailored to each site with as much standardization between sites as possible. Obtain management and workforce commitment to comply, appoint ESD monitors at each site, and periodically audit the sites by off-site authority.

Wall reminders.   Post either the military or commercial ESD-vulnerable signs at each applicable site and add stickers to each workpiece and carrying container (see Figs. 7.1 and 7.2).

How to protect the factory against ESD.   First, it is necessary to measure the sources and amount of static electricity with special electrostatic meters. Next, train the entire company staff in the basics of ESD and its control and further train all those individuals who will have a

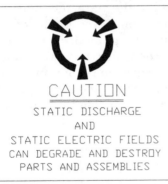

Figure 7.1 Military specification ESD caution sign.

Figure 7.2 Standard ESD caution symbol.

hands-on exposure to production products in the details of proper ESD handling and process control. Finally, incorporate the appropriate materials of different conductivity levels for workstation benches, floors, walls, handling trays, tote bags, and shipping packages. Included in this final item is the work environment itself, the condition of the air, and the condition of general housekeeping.

Electrostatic measurement. It is necessary to initially measure the workplace for static conditions to determine the suitability or unsuitability of materials and work practices in preventing static charge buildup. The source and amount of static charges should be established to serve as the basis of a protection program and as a means of measuring effectiveness of the protection once it is in operation. It is equally necessary to monitor ESD protection products and materials by continuously measuring them over time. Even the best ESD protection devices eventually wear out, and it is essential that deprecation of these protective devices be discovered before ESD damage occurs.

Protective devices and materials should also be measured for effectiveness as part of a qualification program prior to being installed in the user's factory.

Surface resistivity meters. The surface resistivity meter measures the conductivity capability of a material. In this technique of measurement a high voltage is applied to an electrode attached to one point on the surface of the material and measures the current at a grounded electrode. This is a measure of the surface resistivity and is expressed in ohms per square. Surface resistivity measurement is a relatively easy technique but is not always accurate. The surface of material may not always be homogeneous because conduction in an insulating matrix is most often achieved by the inclusion of either uniformly or randomly spaced fillers. In addition, charge migration may be susceptible to local geometric factors and the amount of initial charge distribution and/or proximity to grounded surfaces. Conduction could vary in nonlinear ways, resulting in a false sense of protection or a rejection of an otherwise adequate protection.

A more accurate approach to measuring surface resistivity, when practical, would be to determine the amount of time it takes a material to dissipate a static charge from the surface. By depositing a known charge through a high-voltage corona discharge technique at approximately 2.5 kV, and using a static field meter, making a noncontact charge field measurement of the rate of decay, a more consistent evaluation could be achieved.

Because humidity causes the static charge dissipation measurement technique to vary, it is important to make measurements in the lowest possible levels of humidity likely to arise in practice as well as several widely spaced humidity levels to establish the relative impact of humidity on the effectiveness of the protection.

Static field meter. Statically charged objects have a local electrically charged field around themselves. The amount of charge on the surface of an object can be calculated by measuring the field around it. A simple hand-held induction field meter can provide a compact, low-cost means of measuring the field. Since field strength is a factor of magnitude and distance, measurements need to be taken at reasonably accurate distances from the surface.

Induction static field meters are ideal for static surveys of higher-value electrical fields but are generally insufficient for easy and reliable detection of low value electric fields.

Training. Operator awareness is essential to the success of an ESD program, and proper handling by all others helps reinforce the operator's commitment, especially if management demonstrates their dedication by imposing severe punitive measure on all who violate proper ESD-handling procedures.

All employees should be trained in the basics of ESD prevention and protection. Engineers who work directly with the products on a daily basis in testing, troubleshooting, and repair must know and observe ESD control measures. Quality inspectors likewise have a high hands-on interaction with products and should be trained accordingly. Procurement buyers need to understand the scope and details of the ESD program to impose the proper standards on suppliers. Janitorial workers also need to know the value of proper cleaning and avoidance of casual contact with production hardware and end products.

A comprehensive course should be given to all operators, followed by a brief annual refresher course. A modified course could be given to all other employees to inform them of the necessity and particulars of ESD controls and the potential for causing damage. The training should also describe the devices and materials used to combat ESD.

Housekeeping. All unnecessary plastic items and containers, such as Styrofoam coffee cups, should be eliminated from the workplace in which ESD-sensitive components will be processed. Sweaters and coats should not be placed on the backs of chairs. Floors and benchtops should be cleaned often to remove any insulating films or contaminants that might hold a charge. Maintenance personnel should be supplied with the proper cleaners, waxes, trash-can antistatic liners, antistatic cleaners and wipes, and dissipative floor cleaners. Contract janitorial services should not be expected to select or furnish the proper commodities.

7.7.2   Protection against ESD

The first line of defense against ESD is to prevent static charges from accumulating on nonconducting surfaces and to dissipate charges that have formed without causing damage. Conductive materials are the heart of the defense against ESD. Material conductivity has been categorized into four levels of conductivity by the Electronic Industry Association in their standard number EIA-541 as follows:

1. *ESD shielding* [<1.0 × 10 (4th power) Ω per square]. ESD shields must be capable of attenuating an electrostatic field to prevent damage to or disruption of contained items.

2. *Conductive* [<1.0 × 10 (5th power) Ω per square]. Conductive materials are good electrical conductors.

3. *Static dissipative* [1.0 × 10 (5th power) through 1.0 × 10 (11th power) Ω per square]. When placed in contact with a charged item, this material will dissipate that charge over a given time period.

4. *Insulative* [equal or >1.0 × 10 (12th power) Ω per square]. This material is a poor conductor of electricity.

There are three distinct areas to be protected against ESD: (1) factory facilities, (2) product handling and shipping containers, and (3) people. Factory facilities include environmental controls, flooring, walls, workbenches, handling carts, and production machines. Product handling and shipping includes antistatic bags, tote trays, antistatic tape, antistatic cleaners and wipes, storage containers, bins, and shipping boxes. People-protective devices include wrist straps, dissipative finger cots, antistatic gloves, ESD footwear, smocks, and caps.

Necessary or recommended factory facilities are described in the following subsections.

Humidity. Atmospheric moisture is slightly conductive, and when the workpiece is enveloped in a humid atmosphere (above 40% relative humidity), a thin layer of atmospheric moisture is present at the surface and a gentle, continuous dissipation of static charges occurs. The higher the humidity, the greater the rate of dissipation. Humidifiers are sometimes used in regions where humidity drops below 40%. Humidifiers are generally set to create 50 to 70% relative humidity. Moisture, however, is not considered friendly to electronics, and for that reason humidifiers are not always considered as a prime candidate for ESD protection.

Ionization. Ionized air envelops and neutralizes static charges on insulative materials, which cannot be otherwise removed by grounding, and on surfaces of objects that are not, or cannot, be conveniently protected against ESD. By flooding work areas with a balanced stream of positive and negative ionized air molecules, static charges on the surface of an object—of one or the other polarity—are neutralized by recombining with the oppositely charged air ion. Small benchtop blowers with balanced ionization capabilities are frequently used to supplement other ESD control measures. Installation of a large open-facility ionization system is another method that is sometimes used for larger assembly areas, wave-soldering arenas, and clean-rooms.

There are two major types of ionizers: nuclear and electrical. Nuclear ionizers are expensive and create an annual disposal cost. Electronic ionizers are much more cost-effective and more widely used. Electronic ionizers generate ions by producing a high-voltage differential, approximately 7000 V, between sharp metallic points. As points are dulled by operation, the amount of ions created drops. There is a direct correlation between the number of air ions created (ionized atmosphere increases the conductivity of air) and the static charge dissipation time on surfaces. Imbalanced ionization, which can occur with wear over time, is counterproductive to neutralization and will contribute to the generation of static charges.

Ionization systems should be tailored to match the needs of particular areas to optimize ion levels. Airflow, air conditioning, building configuration, kind of work being done, and humidity control all play a role in altering basic ionization systems. Ion counts for both positive and negative ion levels should be between 5000 and 6000 ions per cubic centimeter.

7.7.3   Removal of ESD by grounding

Static charges are removed by dissipating them to ground. The *floor* is the one surface most constantly contacted by people and objects and is the most convenient surface to neutralize by grounding. Bare concrete surfaces are relatively poor conductors and therefore cannot be depended on for ESD control purposes. Bare concrete needs to be covered by a material with known conductivity characteristics. Vinyl-tiled floor surfaces are nonconducting and also need to be covered with a conductive material.

Mechanical requirements.   There are a myriad of conductive flooring options available from temporary mats to painted coatings and on to permanent tile setting. Floor surfaces are required to resist chemical attack, concentrated mechanical loading, wear, abrasion, and aesthetic deterioration and still continue to function as a conductive dissipator of static charges. Light reflectance, ease of cleaning, and the necessary maintenance requirements are other important considerations in selection of flooring.

Electrical requirements. Electrical flooring performance requirements are centered on the two functions of flooring: (1) to reduce the level of charge generation and (2) to dissipate charges that enter the protected area. Charge dissipation requirements are normally set by the average human body walking routine at 5000 V of charge dissipated to zero in 0.40 s. Resistance measurements are the most commonly used on-site testing. These tests are conducted between two points on the floor surface or a single surface point and a neutral point using electrodes that simulate the human shoe in size and material type.

Surface resistance test methods for on-site floor testing have traditionally used a 500-V operating potential; newer tests now operate at a 100-V potential.

Mats. Grounded, conductive floor mats and runners are the simplest and least expensive equipment to install, are abrasive and chemically resistant, and are easily maintained. This is a very popular floor-protection system in the industry. Mats, which provide only spot protection, have the distinct disadvantage of forcing the operators to be alert to their exact location at all times when handling or carrying ESD-sensitive components or assemblies, as inadvertent misstepping or tripping could be hazardous for the operator and could jeopardize the product. Coverage of large areas with this method can be expensive.

Floor mats can be either rigid, to more readily accommodate wheeled traffic; or soft, to provide maximum operator comfort. Rigid mats are made of fiber-reinforced polymers with conductive additives dispersed throughout the matrix to ensure proper volume and surface conductivity. Soft mats are made from multilayered conductive vinyl that produces a glare-free surface and is easily cleaned. These soft mats, however, have a twofold ($2\times$) charge decay time, in contrast to that for the rigid mats. Corrugated floor mats made of conductive rubber represent another style of soft floor mat that is designed for a workstation where the operator normally stands, such as at a *wave-soldering machine*.

Paint. Conductive floor paint has a low initial cost and is easily applied, but best applied by a specialty team. It is generally applied in combination with a special conductive floor finish. Performance varies with wear and requires frequent maintenance.

Epoxy. Conductive epoxies are durable and seamless and provide a hard surface and reasonably good static protection. Similar to painted surface applications, application of conductive epoxies requires the services of a specialist for proper application. Epoxy floor coverings are also sensitive to existing floor conditions but are long-lasting with good performance.

Dissipative floor finish.   Dissipative floor finish is a low-cost, very effective, existing floor covering made with an acrylic polymer laced with a zinc cross-linked compound that resists voltage generation and functions over a wide humidity range. By itself the surface treatment has a limited performance and requires continual monitoring because of its varied performance with wear. Although initial costs and application are low, its life-cycle costs, due to high maintenance, are high.

Vinyl.   Conductive vinyl floor systems are available in tile form or sheets that can be seam-welded. This floor covering method is vulnerable to wheeled traffic. Initial investment in a vinyl system installation is high but offers permanent protection that covers the entire area of concern. In combination with a top coat of conductive floor finish, this is the best possible floor protection available.

Carpets.   Carpets are available with carbon fiber added to the weave for low-level static protection suitable for office and raised computer room floors. Generally available in tile form, carpeted floors provide comfortable, low noise, and pleasant-appearing protection. Antistatic spray helps reduce the charge generation further, but carpets would not be suitable for higher levels of static protection or factory wear abrasion, chemicals, or wheeled traffic.

Mobile grounding.   Grounding of mobile carts and chairs is accomplished by a small metal drag chain properly attached to the vehicle and draped over a small extended length of the floor. Conductive footwear, such as toe and heel grounders and shoes, are available for the permanent workforce, and temporary heel grounders and booties are available for visitors.

Floor maintenance.   Floor maintenance is critical. Accumulation of dirt or other contaminants on the floor surface create an insulating film interrupting the proper grounding effect of the floor. Where the floor must be damp-mopped to remove stains, water spills, and general grime, special floor cleaners are used to neutralize the surface without affecting the gloss or static dissipative properties. Low-residue wax strippers are used to remove old wax buildup followed by floor rinsing, drying, and application of ESD floor finish to ensure maximum floor finish performance.

Benchtop mats. Benchtop static-controlled mats with conductive additives are used at each workstation as one of the basic ESD control elements.

Utility cart. All metal carts with an attached drag chain to ground the cart to conductive flooring are used to transport ESD-sensitive devices and assemblies.

Bins and shelving. Conductive bins of multiple sizes and designed for ease of handling small parts are available. Versatile, all-metal shelving constructed to allow for ease of reconfiguration is also available for control-room storage.

Product handling and shipping containers. ESD control bags, boxes, and containers are the key elements of this category.

Bags. Protective bags have become an essential part of the ESD control system in the typical electronics production facility. These bags offer excellent protection for individual components and assembly boards; they insulate, shield, conduct charges from the outer surfaces, and present an antistatic surface to induced charges.

Critical applications require carbon-loaded or metalized plastic bags. Antistatic bags have chemically treated outer surfaces or molecularly bonded plastic. Antistats generally rely on atmospheric moisture to properly resist static charge buildup. Treated surfaces are often sensitive to handling, and the protective coverage can be readily disturbed. Generally, bags are designed for one-time usage; however, some can have extended service but should be periodically checked for deterioration. The primary feature for bags is their shielding capability since no electrical field can exist inside a perfectly enclosed conductive covering. Typical static shielding protection levels of materials is 1500 V for antistatic plastic, 7500 V for carbon-loaded plastics, and 15,000 V for metalized laminates.

Transparent bags are available that have metalized layers with abrasion-protection overcoats to protect against induced charges from electrostatic fields and antistatic interior layers to prevent triboelectric generation. These bags are made with an inelastic polyester layer to prevent stretching and puncture. Product identification on the outside of the bag is not necessary because the transparency allows a direct read of the part markings.

Transparent shielding bags are the most common type of ESD protection. These bags can be folded or sealed with a tape or closed with a zip-lock. They are available with padding and with a nontransparent or solid layer of aluminum for the best possible shielding for very sensitive devices or assemblies.

Containers. Containers have all the protective features of bags, except they are sufficiently rigid to resist deformation when handled. Container sizes should be small enough to prevent the protected devices from sliding around, generating static charges. There are conductive tote boxes, metal tote boxes, transparent shielding tubes, conductive plastic tubes, and conductive plastic containers.

Most rigid containers are constructed by vacuum-forcing plastic to prevent extraordinary stresses in the corners.

Caution should be exercised when handling carbon or graphite containers to prevent sloughing, sometimes known as "crayoning," of the outer surface in which abrasion of the carbon or graphite from the outer surface could render the protection ineffective. Sloughing could also release conductive particles in sensitive areas such as cleanrooms.

Use of tote containers. Proper handling procedures of protected products in containers requires that the closed or sealed container or package be placed on a conductive work surface to allow accumulated charges to dissipate from the outer surface. Protected workers can then remove the sensitive device.

People. The electronics workstation is the most critical area in the static control chain because of extraordinarily large damage potential by people, the amount of hands-on touch-labor time, and the amount of nonmetal objects and surfaces. Workstations include incoming matériel receiving and inspection, stockroom and kitting, component placement and soldering, rework and repair, assembly test and system installation, shipping, and field service.

Wrist straps. Wrist straps are the most common and effective means of controlling static and should be worn at all times by anyone touching the components or assemblies. Straps provide a path to ground to constantly dissipate charges before they accumulate.

Wrist straps should be lightweight, comfortable, adjustable, durable, and quick and convenient to hook up. Other styles are also available that include fixed sizes, metal expansion, hook and loop in lieu of the snap, and stretch fabric. For safety reasons it should have built-in current-limiting resistors to protect the wearer from electrical shock and control the static charge dissipation rate. One megohm ($1 \text{ M}\Omega$) is the preferred value of the safety resistance.

TABLE 7.4   ESD Standards

```
MILITARY

1. DOD STANDARD 1686 ·    (CATEGOIZED COMPONENTS INTO
                          THREE LEVELS DEPENDING UPON
2. DOD HANDBOOK 263 ·     ESD VULNERABILITY.)
                         CLASS I     0-1,000 VOLTS
                         CLASS II    1,000-4,000 VOLTS
                         CLASS III   4,000-15,000 VOLTS
                         (CLASS I REQUIRES THE GREATEST
                          AMOUNT OF PROTECTION).
   COMMERCIAL

3. ELECTRONIC INDUSTRY ASSOCIATION STANDARD EIA-541
   FOR PACKAGING MATERIALS.

4. EOS/ESD STANDARD NO.1 FOR WRIST STRAPS.

5. EIARS-471 ESD WARNING LABELS.
```

Garments. Static controlled or cotton outer garment smocks are used as protection to shield normally worn synthetic fabric clothing which easily acquire and hold static charges. For maximum protection, smocks should include crisscrossed conducting threads that are grounded to the wearer's body. Static protection shoes and grounded footwear booties and heel and toe grounders are available. Antistatic gloves and dissipative finger cots are also available.

7.8   ESD Standards

ESD has deservedly received a great deal of attention in recent years from the military and from industry. Table 7.4 lists the standards and guidelines that govern this aspect of the electronics industry.

7.9   Conclusions

ESD damage continues to increase because electronic components continue to become more vulnerable. The level of the workforce's understanding and commitment to the hazards of ESD has not kept up with the level of vulnerability. This is an industrywide, international problem that has been estimated to be responsible for 20 percent of all component failures, and each year the industry has lost ground.

Continuous testing of ESD protection and prevention equipment features should be incorporated as a normal part of the workday and reviewed on weekly and monthly bases. Wrist straps should be measured to determine their proper functioning *while on the personnel.* Testing devices should be readily available and user-friendly. Facilities should be periodically checked using approved ESD testers. Conductive paths have a way of aging with time and need to be periodically tested. Oxidation, rust, and inadvertent maintenance can add paint between joints along critical grounding paths (see Figs. 7.3 and 7.4).

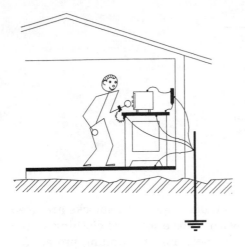

Figure 7.3 ESD direct grounding.

Figure 7.4 ESD indirect series grounding.

The following points should govern ESD follow-through actions:

1. Maintain constant vigilance.
2. Maintain constant training and awareness.
3. Label and mark all susceptible components and assemblies.
4. Follow and maintain proper procedures.
5. Conduct failure analysis.
6. Keep trend analysis data.
7. Conduct periodic audits and reviews.
8. Eliminate all nonconductive, nonshielding workpiece storage containers from the site.

Where ESD control programs have been implemented, analyses show a significant reduction in failures at the production stage and in the field, and return on investments have been calculated between 5 to 1 and 20 to 1 depending on the starting conditions and the scope of the ESD control program. It is clearly one of the best quality investments and business investments management can make.

Bibliography

Bartos, Sara: "Electrostatic Discharge: An Examination of Concerns, Issues, and Solutions," *Electronics,* August 1987, pp. 31–34.

Chubb, Dr. J. N.: "Measurements to Avoid Static Problems with Semiconductors," *Evaluation Engineering,* October 1988, pp. 92–103.

Daniels, Ron: "ESD: A Serious Issue," *Circuit Assembly,* May 1992, pp. 55–57.

Faulkner, Mark: "A Down to Earth Approach to Flooring," *Compliance Engineering,* Summer 1991, pp. 25–31.

Hansel, Ginger, and Michael T. Brandt: "An ESD Primer," *Circuit Assembly,* July 1992, p. 35.

Keeler, Robert: "ESD Control: Making It Work," *Electronic Packaging and Production,* March 1987, pp. 65–67.

Law, S. L.: "A Total Approach to Controlling ESD," *Electronic Packaging and Production,* May 1991, pp. 82–86.

McCraty, Rollin: "Open-Air Ionization: Winning the War Against ESD," *Evaluation Engineering,* March 1983, pp. 78–80.

Plaster, Jeffrey: "Controlling ESD with Partitions," *Electronic Packaging and Production,* November 1990.

Plastic Systems Catalog, "Materials and Systems for Static Control," 1990, selected text.

Staff Writer: "Beware of ESD by Being Aware of ESD," *ADC News,* Fall 1990, pp. 1, 2.

Zalts, J.: "Elements of an Effective ESD Program," *Electronic Manufacturing,* August 1990, pp. 29–32.

Acceptance Testing of SMT

In the past, when some boards had fewer than 500 electrical nodes and assembly functions were relatively simple, testing was relatively simple. Now, SMT assembly functions can exceed 5000 electrical nodes. Today's boards are yesterday's systems. Costs and failure potential of today's boards are far higher than those for yesterday's boards. It does not matter how technologically advanced the board is: if it cannot be delivered as a quality product, it is useless. Testing has consequently become far more complex and far more important for today's boards than it was for yesterday's boards.

Faulty parts need to be revealed as early in the production cycle as possible to prevent adding more value with continued assembly. Testing has become the uncontested and indispensable screen to early detection and correction.

Costs of individual SMT components and individual PWBs can exceed $1,000. Costs of individual assemblies can exceed $20,000. More than ever before, faulty components and faulty PWBs must be prevented from continuing into higher assembly levels and faulty assemblies must be prevented from continuing into the higher system where costs of detection and correction are much higher than at the lower levels of assembly. A $1 fault caught at the lowest level can become a $500 fault when caught at a higher level (see Fig. 8.1). Additional tests have consequentially been introduced into the lower levels of the production cycles. As the SMT components and PWBs become more complex and costly, they are more likely to be electrically and functionally tested at the preassembly level.

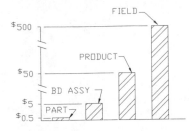

Figure 8.1 Costs to find and correct failure per level of complexity.

8.1 Bare-board Testing

Average SMT bare boards are multilayered with six to eight layers, 2400 or more via holes, 0.200-mm conductor traces, and spaces and land patterns for a mixture of components having 1.270- and 0.635-mm lead pitches. Industry fabrication yields for boards of this complexity have demonstrated the economic need for bare-board testing, and testing even down to the individual inner-layer levels, especially in cases of buried and blind via sequential lamination processing (see Fig. 8.2).

Bare-board testing provides economical payback beyond the fail/pass test at the end of the fabrication cycle. It has become a very valuable window into the PWB fabrication processes and is becoming a primary tool for process optimization. Data gathered by bare-board tests help manufacturing monitor and control processes as well as screen out failed parts. Inner-layer tests can catch short and open circuits and some causes of high-impedance leakage before the faulty layers are processed into higher lamination levels. Bare-board tests provide three main functions:

1. Continuity—verify that each circuit trace begins, ends, and connects to each circuit branch and pad properly.

2. Isolation—-verify that each circuit node does not make contact with any of the other nodes, i.e., ground to power, signal(s).

3. Leakage—verify the absence of high-impedance short circuits.

Tests at the final bare-board fabrication level are especially needed to catch high-impedance shorts caused by the final fabrication processes,

PWB TYPE	TOTAL BDS TESTED	QUANTITY OF FAILED BOARDS	PERCENTAGE OF FAILED BOARDS
DOUBLE SIDED	79,956	11,480	14.4%
MULTILAYER	250,987	53,265	21.2%
INNERLAYER	58,375	5,021	8.6%
TOTALS	389,318	69,766	17.9%

Figure 8.2 Failure rate of commercially produced bare boards.

such as contaminated solder mask, moisture absorption into the laminated layers, and ionic contamination.

8.1.1 Fixtures

Bare-board testers utilize test fixtures consisting of arrays of spring loaded test probes arranged to contact matching test pads on the PWB. These test fixtures, with their arrays of probes, are commonly known as "bed of nails" adapters. The PWB under test, most often referred to as the UUT for unit-under-test, and the bed-of-nails test probes are compressed together using vacuum, pneumatic, or mechanical means to provide adequate electrical contact between the probes and matching PWB test pads. In general, probes should make good electrical contact with pads having average oxidation levels with a compression force of 100 g. Necessary forces in individual situations, however, can vary according to the surface finish and smoothness of the pads. Probes pressed into smooth gold-plated, or leveled solder-coated pads can achieve adequate electrical contact with a 25-g force. Slightly oxidized surfaces could require a compression force of 150 g or higher.

8.1.2 Probe configuration

Test probe sizes and capabilities vary in diameter, length, compression force, test tip head styles, material, finishes, mounting arrangement, and wire attachment. The larger probes, normally used for PWBs, can be arranged on 2.5-mm center-to-center arrays, with a density of 15.5 probes per square centimeter; and the smaller probes on 0.254-mm center-to-center arrays, with a density of 1550 probes per square centimeter. Spring force for the smaller probes is as low as 17 g and as high as 190 g for the larger probes. The bullet-shaped probe head style is ideal for SMT boards.

Newer probes have spiral plunger shafts that cause the contact tip to begin rotating as it makes initial contact with the test pad and continues to rotate to a maximum of 90° by completion of the full plunger travel. By combining this rotational motion with the probe's spring force and using a chisel head style with sharp edges, test pad surface contamination can be penetrated at lower compression forces. Rotating probes are available in the 2.54- and 1.90-mm center-to-center sizes.

Test probes are inserted into probe receptacles with a press-fit and the receptacles are inserted into a fixture baseplate with a fixed mounting that consists of either adhesive or mechanical fastening. Receptacle wire termination options include crimp contacts, wire solder cups, wire-wrap posts, or male connector pins. Basic probe construction consists of a barrel housing, spring, and plunger with test head. The probe

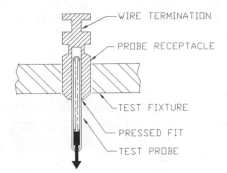

WIRE TERMINATION

PROBE RECEPTACLE

TEST FIXTURE

PRESSED FIT

TEST PROBE

Figure 8.3 Test probe-fixture interface.

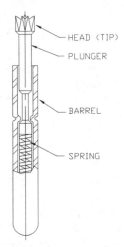

HEAD (TIP)

PLUNGER

BARREL

SPRING

Figure 8.4 Test probe.

barrel is sized to be press-fitted into the receptacle with the electrical and mechanical connection between the two being dependent on the press-fit (see Fig. 8.3). This arrangement allows ease of reconfiguring the probe test pattern arrays to fit individual PWB patterns and ease of replacement of faulty probes without rewiring the fixture (see Fig. 8.4).

8.1.3 Fixture configuration

Test fixtures can be either fixed pattern matrixes or universal matrixes. Universal test fixtures consist of a full complement of wired receptacles arranged in a standard grid matrix. Probes are inserted as needed to match specific UUTs. Translator plates are provided within the fixture to guide the probes in the event the UUT test pads are offset from the universal grid matrix. Under these conditions the probes are tilted, diminishing the compression force and, partly as a result of probe deflections, reducing accuracy of registration. This becomes exceedingly difficult when the amount of compression force, dictating the structural size of the probe, must be compromised to accommodate fine-pitched 0.508-mm test pad patterns.

Mixed center testing fixtures, where large, medium, and small probes are all comounted to a single fixture, are becoming more common in response to SMT assemblies that are becoming increasingly populated with 1.27-mm, 0.63-mm (and smaller pitch) leaded components and routed with high-density conductors. Such fixtures can cost $3,000 and more.

8.1.4 Probe pressure

Further fixture compromise is needed when the compression force needed on the UUT exceeds the board's ability to withstand that force or when the fixture's ability to deliver the necessary force is exceeded, especially in the case of vacuum-applied force. When individual probe compression force is 100 g, the overall force for a board involving 6000 electrical nodes would equal 600 kg and could exceed the limits of atmospheric pressure.

8.1.5 Moving-probe tester

An alternate bare-board tester that does not use multiple probes uses two computer-controlled moving probes that can access any two test pads and test for shorts, opens, and high-impedance shorts (leakage). This method can test two pads spaced less than 0.508 mm apart and test multiple nodes at a rate of 350 nodes per minute.

Moving-probe test systems are ideal for high-density boards, development work, limited production quantities, and flexible production mix of assembly configurations. There is little to no changeover down time and no storage of test fixture adapter boxes.

8.1.6 Bare-board test voltage

Bare-board test voltages vary between equipment suppliers and between users. Commercial boards are tested at 10, 20, and 40 V dc at 20 Ω. Military boards are tested at 275 ± 25 V at 10 Ω.

8.1.7 Test equipment and environment

Testing should be done in a clean, dry, dust-free, temperature- and humidity-controlled environment. Well-regulated electricity should be used for primary power. Common tester hardware features include

TEST SYSTEM	ANALYZER COST ($1,000)	PIN/FIXTURE COST
DEDICATED FIXTURE	50	$5/PIN=$12,000
UNIVERSAL FIXTURE	500	$1/PIN=$2,400
MOVING PROBE	90	NONE

Figure 8.5 Moving probe versus fixture test cost.

TEST SYSTEM	SET-UP TIME (MIN.)	THROUGHPUT
DEDICATED FIXTURE	30	3 BDS/MIN.
UNIVERSAL FIXTURE	60	3 BDS/MIN.
MOVING PROBE	5	300 POINTS/MIN.

Figure 8.6 Moving probe versus fixture test time.

solid-state switch gate logic, power supplies that can supply test voltages of 10, 20, 40, and 300 V, compression mechanisms capable of supplying adequate contact force to meet the needs of all UUTs, a computer that includes interfacing for downloading CAD files and network activities fed by an uninterruptible power supply, self-learn programming from a golden board, a test throughput measured in minutes per boards, a good-board/bad-board testing accuracy with self-test measurement of voltage and resistance parameters, low downtime for repair and maintenance, user-friendliness, well-documented service manuals, and factory service on-call.

Bare-board testers and moving probe testers normally cost between $100,000 and $200,000. Test fixture costs, however, are eliminated with moving-probe testers (see Fig. 8.5). Operational throughput comparison between bed of nails and moving probes is shown in Fig. 8.6.

8.2 High-Potential Testing

High potential testing verifies *nonelectrical* continuity between power and ground planes in PWBs. This test is performed to detect three types of defects:

1. Resin voids

2. Shorts between power and ground planes

3. Inadequate clearance between power and ground planes.

To perform these tests, a high-voltage potential is applied in incrementally increasing voltage levels applied between the PWB power and ground layers from 50 to 1500 V. High voltage is held for 10 s to allow time for arcing or current leakage above 10 μA to interrupt the high voltage, revealing a failure.

Laminate voids (gas pockets buried within the PWB) is one cause for voltage breakdown in PWBs. Ionization of the entrapped gas is initiated by the high voltages, resulting in arcing or current leakage. Innerlayer misregistration shifts can position hole walls close enough to traces to arc under high-voltage testing. Outerlayer high-impedance contamination, described in Sec. 8.1, can also be detected by high-potential testing. Underetch and overetch also affect clearances and can result in arcing. Hot-air solder leveling can leave partial solder webbing between lines, holes, and pads. Microscopic slivers can be present as a result of partial processing or poor cleaning. High potential testing has proven to be a good failure analysis tool in fault location and determining probable cause of failures.

8.3 Surface Insulation Resistance Testing

Surface insulation resistance (SIR) is a measure of an insulation material's resistance to electronic current flow along its surface. Testing for SIR is a practical way to test for surface cleanliness (see Fig. 8.7). With the recent introduction of no-clean and water-soluble fluxes and the introduction of cleaning substitutes, PWB surface cleanliness has become an issue of concern. SIR testing is specified in IPC-SF-818 for the general industry and required by the military as specified in MIL-F-14256E. This issue is particularly important for SMT PWBs that have fine-pitch components and tight circuit densities. Degradation of surface integrity can have a great effect on the circuit in complex assemblies having high-density components and spacings.

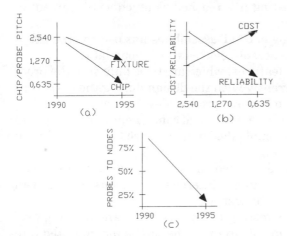

Figure 8.7 Test and fixture trends.

SIR valves are determined by the use of SIR test coupons that have a comb pattern described in IPC-B-25. These coupons are used to determine the SIR values of the basic material, but more importantly, they are also used to measure the effective results of fluxes, processes, and cleaning fluids over a given time period with a 50-V bias applied. Resistance levels of 100 MΩ are normal for clean boards, although levels of up to 1000 MΩ have been suggested for specification purposes.

8.4   Loaded-Board (Assembly) Testing

8.4.1   First role of testing

Board assembly testing is being called on to serve two dual-purpose roles. The first role is to perform functional acceptance testing by accessing all circuit nodes and sequentially verifying the functional parameters of each component and verifying as much of the total circuit performance functions as is economically feasible. This first role is geared to screening all assemblies with the primary goal of not allowing faulty products from proceeding any further. A second goal, in this first role, is to isolate faults for ease of repair.

It is becoming increasingly complex and difficult to completely test and fault-isolate failures on board assemblies, particularly if the assembly includes advanced VLSI and ASIC (very-large-scale and application specific integrated circuit) chips, fine-pitched and high-density packages, multichip modules, and TAB terminated chip-on-board (COB) devices. For the same reasons it is becoming more and more complex and difficult to mechanically interface with these assemblies for test purposes (see Fig. 8.8). Testing is representing a larger portion of assembly costs because of these complexities. On some very complex, high-density assemblies testing has reached as much as 50 percent of the total costs.

Assembly test engineering of SMT assemblies has become the art of accurately determining the accept/reject status by partial test.

Test philosophy and test features, which are to be built into the product, are established concurrently with the design during the early conceptual formalization period for new product development.

Test pads, test connectors, and additional component parameters are integrated into the basic assembly design to improve functional testing and diagnostic troubleshooting.

Bed-of-nail test fixtures, described in Sec. 8.1, are used to access widespread nodal points. Functional interface connectors and built-in test connectors are also used for testing.

Extra gates to control the testing of clock circuits, and added gates, switches, and jumpers to control circuit paths can be incorporated into

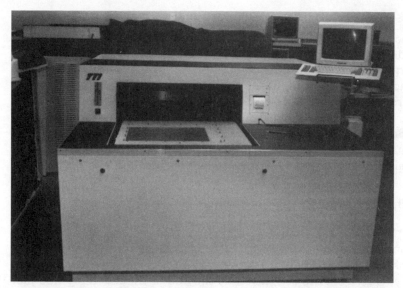

Figure 8.8 Bare-board test machine planar resistor testing. (*Courtesy of Test Technology International, Inc.*)

the assembly circuits. Means can be added to initialize registers, flip-flops, counters, and solid-state devices. Built-in tests for microprocessor-based chip sets, when appropriate, can be added. Resistors can be added for all unused inputs to isolate individual devices and allow circuits to be back-driven by the test set.

Signals can be routed to edge connectors using built-in test logic. When electrically and mechanically possible, route circuit nodes to a single test site, preferably to a location conveniently reached by a multiple test probe apparatus. Attach individual test pads to each component lead pad for maximum testability of circuit functions, debug test, troubleshooting test, and failure isolation; otherwise use at least one test pad per circuit node. Edge-board finger connectors can be used as test connectors, even for military applications. Test functions can sometimes be routed to otherwise unused pins on functional connectors. Snap-off coupon-type factory test aprons can also be used when space is at a premium, such as with very small boards. When total nodal access is not possible, or otherwise not desirable, test-access caps (small test fixtures that fit over the top of individual components with the appearance of top-mounted sockets) are used to gain access for tests.

Caution. Do not permit the body of the solder joints, the edge of via holes, or the conductor areas of devices to be probed. Pointed probes, even under gentle hand pressure, can initiate failures or disguise an intermittent failure in the joint during troubleshooting.

8.4.2 Second role of testing

The second of the dual-purpose roles for loaded-board testing is a recent addition to the first. In this second role measurements and tests are performed for the purpose of monitoring and reporting the status of fabrication processes and, in the event of failures, help identify the errant processes that contributed to the failure. With this second test role, it is envisioned that processes will progressively improve after each failure feedback from test and each corrective action cycle. Eventually, the processes should be corrected enough to make it possible to eliminate the first role, the product verification role.

8.5 Environmental Stress Screening

8.5.1 ESS purpose

Environmental stress screening (ESS) is a manufacturing test to verify workmanship and not an engineering test to measure design or reliability. ESS tests are conducted at the end of the assembly line just prior to functional acceptance testing on 100 percent of completed products to detect latent manufacturing defects related to the assembly and its components. ESS testing has proved to be a cost-effective step for many commercial products used in nonbenign environments or otherwise subject to severe shipping and handling and is a contractual requirement for all military electronic products. Latent manufacturing workmanship defects induced failures encountered by the user have been dramatically reduced by ESS testing.

8.5.2 ESS initiated by NASA

ESS testing was first introduced by NASA as one of the measures taken to ensure that the safest possible systems were used for early manned space flight. The military quickly adopted NASA's ESS test philosophy and methods. This form of product testing has evolved over the years and matured to the point of industrywide acceptability.

8.5.3 Selected environments

The aim of ESS is to vigorously stimulate the assembly enough to precipitate latent defects but not enough to diminish useful life of the product. Various tests were attempted over the years in search of the ideal, cost-effective test(s). Many have proved to be ineffectual, costly, or detrimental. Altitude, acceleration, humidity, mechanical shock, vibration, thermal shock, temperature cycling, and electrical stress are many of the tests tried. Two tests in combination with one another have proved to be effective: temperature cycling and random vibration.

8.5.4 Temperature cycling

Temperature changes produce a structural reaction in the assembly materials in the form of CTE mismatch and thermal conductivity variations. These material reactions produce the stresses that reveal weaknesses. High and low temperature extremes, the number of cycles, and the rate of temperature change are three of the key factors that determine the effectiveness of ESS testing. Each product should be reviewed to determine the ESS temperature profile that matches it. Military board assemblies, used in a ground launch missile, have been successfully ESS-tested at −55 to +60°C at a temperature change rate of 20°C/min for a total of six cycles. Complex assemblies with more than 400 components are cycled 10 times and simple assemblies with 100 components, one to three cycles.

8.5.5 Random vibration

During the earlier years ESS vibration consisted of sine-wave testing to a specified spectrum of vibration frequencies and vibration amplitudes that were set somewhat higher than that expected in service.

Experience and analysis have shown that actual vibrations for airborne and ground-based vehicles contain a wide band of randomly changing frequencies, a vibration spectrum quite different from that used for sine-wave tests. Sine-wave vibrational energy is concentrated in one frequency at a time and can cause failures that would not occur in service. Sine-wave testing can also miss detecting latent defects. When sine-wave shaker test frequencies coincide with the natural frequency of the product, the vibration amplitude of the product can be far higher than intended, resulting in nonservice-related test-induced failures. Random vibrations, being closely matched to service vibration, rarely induce nonservice related failures in properly designed products. Products that have passed random vibration rarely fail in service. Workmanship flaws missed by sine-wave testing are revealed by random vibration.

Military products are randomly vibrated, in accordance with NAVMAT P-9492, at 6-g rms (root mean square) over a 20- to 2000-Hz frequency band for 10-min periods for a single axis and 5 min more for each additional axis.

8.5.6 Combination testing

Combining temperature cycling with random vibration is, by far, the most effective method of ESS testing. Single machines that combine the shaker within the temperature chamber cost $200,000 and more. Because of throughput, multiple machines are often necessary. ESS testing is necessary to screen products that are used in nonbenign environments, where user safety is involved and where warranties of complex, high-priced products are involved. ESS testing is not necessary for high-volume, low-cost consumer products or where production yields are very high with little chance of latent manufacturing workmanship defects.

Bibliography

Baker, Jess: "Automated Bare-Board Solderability Testing," *PC FAB*, December 1991, pp. 42–43.

C.A.D.—Its Value in Bare Board Testing, Testerion, Inc. handout publication.

Conti, Joseph: "Bare Board Test Methods for High Density Board Designs," *Electri-Onics Electronic Edition*, September 1987, pp. 31–32 (part 1); October 1987, p. 39 (part 2).

Engelmaier, Werner: "Environmental Stress Screening and Use Environments—Their Impact on Solder Joint and Plated-Through-Hole Reliability," *IEPS Journal*, 1991, pp. 7–13.

Fennimore, Jack: "Martin Marietta's Approach to SMT Standards," *Journal of SMT Standards*, Spring 1988.

Hroundas, G.: "Economics of Bare Printed Circuit Board Testing," *T.C. Connections*, vol. 2, no. 3, July 1986, pp. 1–2.

Levine, Bernard: "SM Resistors Also Sought for Miniaturized Products," *Electronic News*, June 10, 1991, pp. 16–18.

Lyman, Jerry: "Components for SMA Arrive," *Electronic Week*, April 8, 1985, pp. 49–53.

"Ostby-Barton Spring Loaded Test Probes," *Ostby-Barton Test Probes Product Catalog and Guidelines.*

Reynolds, C. Edward: "SM Connectors Push toward Higher Density," *Electronic Engineering Times*, October 19, 1987, pp. T26–T28.

"Solder Pastes," *Alpha Metals, Inc. Product Catalog and Guidelines*, 1986.

Spitz, S. Leonard,: "Process Control Begins with Testing the Bare Board," *Electronic Packaging and Production*, December 1988, pp. 46–49.

Werther, Bill: "Adapting SMT for Through Hole Assembly," *Surface Mount Technology*, March 1992, pp. 41–43.

Young, Bret: "Mapping the Labyrinth of SMT Packaging," *Electronic Engineering Times*, October 19, 1987, p. T24.

Rework and Repair

9.1 Definitions

Repair, by definition, is performed for one of the following reasons:

1. To remove and replace a defective component

2. To correct an assembly process error

3. To fix and restore a damaged board

Rework, by definition, is performed to alter the board and assembly to either (1) correct a design problem or (2) effect an engineering or system change. The processes used for repair are also used for rework. The two, for all practical purposes, are synonymous.

It is possible for SMT articles to be repaired and restored to the same level of reliability and quality as first-run articles. However, it is not easily done, and only the best operators should be assigned to perform the work.

9.2 When and When Not to Repair

Rejected SMT boards and assemblies fall into one of two categories. Either they are defective and will not function properly, or they are out-of-tolerance, indicating a process that is beginning to go out of control. If they are defective, as bridging solder or missing solder, they must be repaired. If they are process indicator errors, as uncentered components or excessive solder, they should be accepted as is, without touching them. The errant process, of course, would need immediate attention. Assemblies requiring design corrections, or system changes, should be reworked.

9.2.1 Touch-up

Not all SMT solder joints are improved by touch-up. If a joint has marginal quality, it is often better to accept it as is than to add or subtract solder or improve its appearance. Touch-up can be detrimental to the joint's reliability.

9.2.2 Cautionary notes

Conventional rework processes using cored solder wire and pencil-type soldering irons should be used cautiously. Some passive chip components have a tendency to attach themselves to the soldering iron tip because of their small size and ultralight weight. Rosin flux, in core solder, begins to char and decompose at 285°C and, once charred, is very difficult to clean. Charred residue can be conductive and corrosive. Established repair procedures for IMT may need to be revised when repairing IMT components on mixed IMT-SMT assemblies (Type II or III) to protect adjacent SMT components from mechanical or thermal damage.

Repairing an SMT component on one side of a double-sided assembly may inadvertently affect component solder joints on the opposite side, especially on thin boards and in the close proximity of PTH vias.

SMT repair and rework should be avoided if possible. It is a tedious, time-consuming operation requiring the best-skilled operators using unique tools. It is costly and prone to error, but, because of costs and, very often, delivery schedules, there is no other choice but to repair or rework.

9.3   Process Control

SMT is a technology based on process controls and operational consistencies. It is very difficult, but necessary, to impose these same disciplines on the rework and repair arenas. All the considerations and controls that go into the fabrication of the original articles need to be incorporated, in one way or another, into the rework and repair operations. The difficulty is transposing automation process controls to manual or semimanual processes.

Control of heat flow needs to be as much at the center of rework process as it is for production processes. Time-temperature profiles, established to protect temperature-shock-sensitive components and to optimize flux and solder dynamics by promoting proper heat flow into solder joint terminals, must be duplicated for the same reasons. There is, however, one large difference between production soldering operations and rework and repair operations. Production operations are geared to mass soldering, whereas rework operations are geared to single components in confined spaces.

Adjacent components must be kept below 150°C to avoid degradation and possible failure of those components or their solder joints. It is not easy to desolder a FP component with 196 leads, pouring in enough heat to simultaneously reflow all the joints without promoting intermetallic growth or affecting neighboring components. Nor is it easy to resolder the replacing component. Often it is necessary to have two procedures, one to desolder and one to resolder, with two different sets of tools and machine aids, using two different soldering methods.

9.4   Basic Methods

Basic heat sources used for original soldering are also used for rework soldering: conductive (soldering irons, hot plates, hot bars), convective [vapor-phase (VP), hot-air, and hot-gas soldering], and radiated (IR and laser soldering).

There are three types of rework and repair adapters: (1) single-point (one contact at a time), (2) multiple-points, and (3) entire-components. Contact tool bits can be either tinned or nontinned. Tinned bits (tips) transfer solder as well as heat; nontinned bits, which transfer heat only, are ideal for solder paste applications.

Source heating can be delivered either continuously or pulsed. Of the two, pulse heating is best, but each pulse must deliver enough power to supply the total heat load and be able to recover for repetitious operations. Delivery of forced hot air, or gas, to a particular component is relatively simple. Controlling the exhaust from that forced hot air, or gas, to prevent lateral heating of adjacent components, is not so simple.

Component handling is accomplished by using tweezers or vacuum picks. Vacuum picks are the better choice. They present little to no damage to the component. Premature lifting by tweezers of components in desoldering operations can damage the components and can impose enough force to lift the PWB pads.

The most difficult part of the contact heat flow method is achieving intimate, thermal contact with all the component leads simultaneously for just long enough time, and no longer, needed for solder to reflow. Contact tips adapted with variable adjustments are used to compensate for variations in component lead geometries (see Figs. 9.1, 9.2, and 9.3).

It is difficult, but possible, to remove components that have been adhesively attached to the PWB. By applying localized heat to soften the adhesive, components can be plied free after the desoldering operations.

Techniques used to remove conformal coating from SMT assemblies are not unlike those used for repair of IMT assemblies. General coat-

Figure 9.1 Manual, contact, removal tool for PLCC and LLCC components. (*Courtesy of Pace, Inc.*)

ings can be removed chemically, by thermal parting, mechanical abrasion, or hot-jet application. Any one, or a combination of these, can remove any of the coating normally used for electronics. Difficulty comes when the coating runs under the component during initial assembly and acts like an adhesive joint during removal operations. Special thermal parting tool tips are often necessary to apply the heat directly to the sandwiched coating, or adhesive, once the solder has been removed.

9.5   Rework Configuration

9.5.1   Add-on leads

One of the prominent reasons for design changes is to compensate for unexpected thermal variations associated with CTE mismatch be-

Figure 9.2 Manual, contact, removal tool for flat packs. (*Courtesy of Pace, Inc.*)

tween the PWB and the leadless components. Another cause is in response to the lack of leaded component availability. Both of these problems can be solved by adding adaptive leads to existing leadless chip carriers. Add-on leads, mentioned in Chap. 3, are an ideal solution for component design changes in the midst of production. Add-on leads should not require a change in the PWB land patterns, nor should they require a change in the component. Leads should be simply added on via soldering or thermal compression bonding without causing changes or physical damage to the component body or its terminations. Basic reliability of the component should not be degraded in any way by the lead attachment operation.

Industry has produced three basic add-on lead configurations: (1) rectangular spring-shaped leads soldered to the top or bottom of the component, (2) rectangular spiral-shaped copper columns soldered to

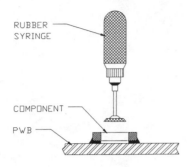

RUBBER
SYRINGE

COMPONENT

PWB

Figure 9.3 Hand-held vacuum pickup tool.

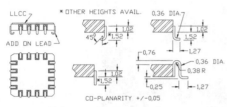

Figure 9.4 LLCC add-on leads.

the bottom of the component (see Fig. 3.4), and (3) round copper leads shaped as J, gull-wing, or I leads thermocompression-bonded to the castellation cavities of the leadless component (see Fig. 9.4).

9.5.2 Add-on components

Additional components can be incorporated safely and reliably into the assembly without first making artwork changes to the PWB. SMT is much more versatile and conducive to retrofitting with reconfiguration changes via rework than is IMT.

Passive devices can be bonded to the PWB and then wired into the assembly by terminations to either the PWB or to other components (see Fig. 9.5). Active devices can also be bonded and wired to the base assembly. Figure 9.6 shows a LLCC flipped ("dead cockroach") and interconnected.

9.5.3 Circuit routing alterations

Existing PWB circuit conductors can be severed or have their PTHs drilled out to effect a circuit disruption (see Fig. 9.7). Rerouting circuit paths and node hookups can be achieved by adding discrete wires (see Fig. 9.8).

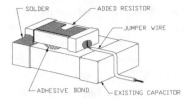

Figure 9.5 Add-on, rework resistor.

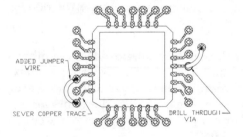

Figure 9.6 Rework with flip-component.

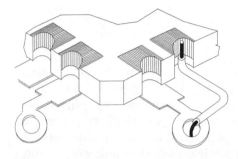

Figure 9.7 Rework features for LLCC device circuit.

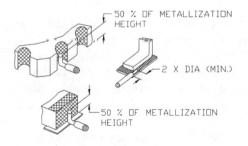

Figure 9.8 Discrete rework wiring.

9.6 Typical Repair Operation

9.6.1 General observation

When replacing a component on ceramic substrate with hot gas, the entire board is first heated to 120 to 150°C, and then the hot-gas nozzle is located over the component to be replaced and the reflow cycle is started. As soon as the solder reflows, a vacuum pick lifts the component with a slight twisting motion to break the solder surface tension and lift cleanly. Next, determine whether there is enough solder left on the pad for the new component. Required amounts of additional solder paste can be added by using a syringe. Castellations on LLCCs should be pretinned with solder prior to attachment. The new component is placed and the solder is reflowed using the hot gas. A bake-out of the solder paste, prior to reflow, is seldom required because of the small amount of solder usually added and because of the controlled ramp-up of the heat.

9.6.2 Rework soldering methods

Hot-air or hot-gas method. Hot air or hot gas is the predominant method for SMT rework and repair. This method is rapid and heats diverse-shaped and -sized parts evenly with noncontact localized heating. It is a clean, inexpensive method but is difficult to control and consequentially requires specialized nozzles for just about every part geometry. Rework machines can be purchased, depending on complexity, from $2,000 to $20,000 for manual machines and up to $100,000 for fully automated machines.

Hot plate. Hot plates are used for flat-surface, single-sided assemblies having no heat sinks or embedded metal layers. This is a simple, inexpensive technique that does not require specialized tools, hoods, or shields. The disadvantage of this method is that all solder joints are reflowed, regardless of how limited or simple the repair is. Hot plates are available for $100 and cost up to $5,000.

Vapor phase. Vapor phase (VP) is used for reattachment. If rework and repair are extensive, involving a number of components, this approach may be justified; otherwise its only advantage is automatic process control over the reapplied solder joints. A disadvantage is that all joints are reflowed. Existing equipment and processes, however, can be used.

Infrared. IR soldering, like VP, is used for reattachment. With specialized machines, tools, and shieldings, localized heating can be accomplished, eliminating the reflow of all solder joints. Cost ranges from $10,000 to $50,000 for machines and adapter tools.

Lasers. Lasers can be used with precision and highly localized controlled heating. The only major disadvantage is equipment cost, if machines do not already exist for initial production. Laser machines start at $100,000.

9.6.3   Component replacement

Passive devices.   To *remove,* flux area of removal, use specially designed solder irons, wick solder off (light touch, no pressure), and then clean.

To *reattach,* tin component terminals if not done previously, position component, flux pads and component terminals, solder one side and then the other, and then clean.

LLCCs.   To *remove,* flux area of removal, use specially designed hood or iron tips, reflux if required and remove excess solder, and then clean.

To *reattach,* tin component leads if not done previously; flux PWB pads and preapply; position component, shim to proper height, flux, and tack-solder two leads; flux and solder individual leads or place solder preforms, or paste, and reflow solder; and finally clean.

Leaded components.   To *remove,* apply flux to leads, heat and lift each lead if by iron or entire component if otherwise (use slight rotational motion as solder reflows to break solder surface tension and allow a clean lift), wick solder off if component is adhesively bonded and then deal with adhesive in accordance with vendor recommendations or company procedures (use ultralow pressure when wicking), flux and wick pads clean after component removal, then finally clean.

To *reattach,* tin component leads if not done previously; flux pads and reapply proper amount of additional solder (preforms, paste); position component, flux, and tack-solder two leads; flux and solder individual leads or reflow entire component with hot-air or hot-gas, vapor-phase, IR, laser, or hot-bar method.

9.7   Rework and Repair Problems

The following is a brief list of generalized problems associated with rework and repair:

1. Burning PWB surface
2. Lifting PWB traces and/or pads
3. Component damage
4. Thermal control for adjacent components
5. Miniature components
6. Component assembly density
7. Conformal coating removal
8. Adhesive bonding beneath component

9. Solder mask

10. Diverse standard and FP lead configurations

9.8 Repair Station

Ideally, repair tools that deliver heat should have the following features or aids:

1. Adjustable flow control

2. Adjustable temperature control

3. Nozzle and tip temperature feedback sensor

4. Adjustable timer

5. Quickly interchangeable nozzle and tip or adapter when hot or cold

6. Adjustable and movable illumination

7. Zoom stereo microscope

See Fig. 9.9 for a process-controlled repair station and Fig. 9.10 for a precision-focused hot-gas system.

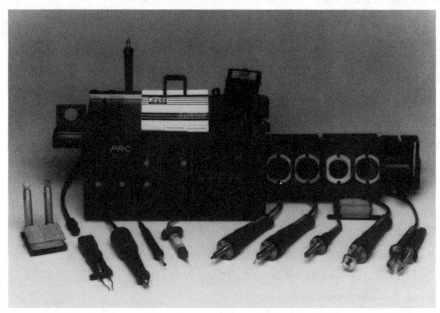

Figure 9.9 Process-controlled manual component and repair station. (*Courtesy of Pace, Inc.*)

Figure 9.10 Precision-focused hot-gas system. (*Courtesy of Pace, Inc.*)

9.9 Electrostatic Discharge

All normal ESD precautions and routines should be followed as rigorously in the rework stage as on the production line.

Bibliography

Ablagnaro, Lou: "Safe Repair of Surface Mount Assemblies," *Electronic Packaging and Production*, November 1990, pp. 55–61.

Capillo, Carmen: *Surface Mount Technology: Materials, Processes, and Equipment*, McGraw-Hill, New York, 1990, pp. 183–203.

Floeter, Ed: "A Practical Approach to the Repair of Surface Mounted Assemblies," *Hybrid Circuit Technology*, October 1986, pp. 37–39.

Hodson, Timothy L.: "Solving Rework and Repair Problems," *Electronic Packaging and Production*, January 1992, pp. 21–26.

Martel, Michael L.: "The Use of Collimated Infrared in Soldering SMT," *Surface Mount Technology,* November 1990, pp. 60–62.

Parton, Jay W.: "Conformal Coating Removal," *Circuits Assembly,* November 1991, pp. 28–33.

"SMT Repair Methods," from MIL-C-28809, *Circuit Card Assemblies: Rigid, Flexible, and Rigid-Flex.*

Snadeau, Renée F.: "Hermetic Chip Carrier Assembly Modifications," *Proceedings of Technical Program, International Electronic Packaging Society,* November 1982.

Solberg, Vern: "Sub-Module Corrects Deficient Assembly During Production," *Surface Mount Technology,* June 1988, pp. 17–19.

Weightman, Neil: "Microabrasive Blasting in Electronics Applications," *Circuits Assembly,* November 1991, pp. 36–39.

Advanced SMT Trends— Now and in the Future

For electronics, the second industrial revolution began with the introduction of solid-state semiconductor devices in the early 1950s. Continuous improvements in hardware size and circuit functions have been occurring at an increasingly accelerated pace ever since. The switch from IMT to SMT was inevitable. The switch from wire bond to TAB and from packaged chips to unpackaged chips will also be inevitable. This book presents a snapshot of the SMT industry taken in 1992 during this critical period of ongoing evolution from IMT to SMT and beyond (see Fig. 10.1).

Regardless of a dim future, IMT is, at this time, still used in the majority of the established electronic products. SMT, however, is coming on strong and will soon replace IMT as the dominant packaging approach. In spite of SMT benefits, and soon-to-be dominance, there is still a reluctance on the part of some companies to convert to SMT. As soon as the emerging SMT processes and producers assume dominance within the industry, the newer evolutionary technologies already appearing in some products and presented as trends in this chapter will soon thereafter become the strong contenders for dominance in the industry.

TRENDS	
HIGHER DENSITIES	REDUCED I/O PITCH
HIGHER POWER DENSITIES	MATCHED CTE MATERIALS
HIGHER SPEEDS	STANDARDIZATION
HIGHER I/O	REPAIRABILITY
INCREASED LEAD COUNT	LOWER VOLTAGE
LARGER CAVITIES	INCREASED CROSSTALK
LARGER CHIPS	REDUCED NOISE MARGINS

Figure 10.1 Electronic packaging trends driving future designs.

10.1   Multichip Modules

Multichip modules (MCMs) are the product of a subtechnology that has evolved from the hybrid industry to match the needs of SMT. Active and passive chips with thin-film and thick-film passive circuit elements are combined in a single, non–hermetically sealed or hermetically sealed package to form specific circuit functions that require a specialized packaging configuration geared to meet unique circuit characteristics, such as high speed and higher-density system packaging requirements.

10.1.1   Comparison of MCM to hybrids

Multichip modules are not unlike earlier hybrid modules. The likeness between MCMs and hybrids has caused some confusion within the industry. The following table lists features which show the difference between the two.

MCM	Hybrid
Large package (>50-mm side)	Small package
High I/O (<1.27-mm pitch)	Low I/O (usually 2.54 mm)
Large chips (>6-mm side)	Small chips
High-I/O chips (>100)	Low-I/O chips
High-density substrates (<0.05 mm)	Density (>0.127 mm)
Chip area per substrate (>40%)	Low-density chip area per substrate
Chips per substrate (>12)	Chips per substrate (>12)
Pretestable TAB chips	Nonpretested chips (wire-bonded)

10.1.2   Benefits and challenges of MCM

Use of MCMs in many applications reduce volume and weight, improves circuit performance, reduces number of parts to be assembled, reduces interconnections at the PWB level, and increases reliability. The challenges facing MCM are the thermal management of higher-power densities, ground bounce with high-speed chips, precision repair, and the development of newer technologies, in many instances, that are involved with bare chip assembly, manufacturing yields, tests, and repair.

10.1.3   MCM categories

MCMs are subdivided into three categories according to their substrate materials and construction techniques: MCM-C (total ceramic), MCM-D (dielectric on ceramic or silicon), and MCM-L (organic laminate).

MCM-C substrates are cofired multilayer ceramics using either high-temperature, low-temperature, or photoceramic processes.

MCM-D substrates are constructed by using silicon, ceramic, copper, or metal composites as the basic substrate that can include ground and power planes and signal conductors. Onto this basic substrate is added signal and ground conductors by sequentially building up layers of unreinforced dielectric materials intermixed with aluminum, copper, or gold conductors and vias added by fine-line lithography and sputter or plated thin-film metalization processes. Conductors are 10 μm wide, 5 μm thick, and 24 μm-diameter via holes through 10-μm-thick photosensitive polyamide.

MCM-L substrates are fabricated of conventual, and/or newly developed reinforced or unreinforced organic materials using conventional substrative, semiadditive, or fully additive processes normally used for PWB fabrication.

MCM substrates are much more complex and larger than their earlier counterpart hybrid substrates. One major challenge facing MCM is the cost of the substrates (see Fig. 10.2). Without a dramatic reduction in substrate costs, MCM-D usage will remain at its present low level.

10.1.4 MCM substrates

Substrates are at once the heart and challenge of MCM and the basis for the three MCM categories. With their small, compact sizes, substrates can be fabricated to meet the subminiaturization required for high-speed circuit performance and packaged to protect the cluster of otherwise bare chips. Technically, today's substrates have met the needs of MCMs for current requirements. The challenge now is the cost of the high-density MCM-D substrates. Aside from the potential cost due to untested bare chip yields at final MCM assembly, MCM-D costs are driven by the unusually high cost of the subminiaturized, high-performance substrates. Substrate costs for MCM-C and MCM L are high but manageable.

MCM-C. Conventual ceramic materials and methods are generally used for MCM-C substrates.

	MCM-C (CERAMIC)		MCM-D (DIELECTRIC)		MCM-L (LAMINATED)
SUBSTRATE	COFIRED	LOW-K	SIL/SIL	LOW-K ON CERAMIC	PWB
LINE DEN. (cn/cm²/layer)	20	40	400	200	30
LIN. WIDTH/SPACE (mm)	125/125-375	125/125-375	10/10-30	15-25/35-75	750/2250
MAX. SUB'S SIZE (cm)	10-15	15	10	10	66
DIEL. CONST.	9.0	5.0	4.0	2.4-4.0	4.5-5.0
I/O/cn ———— PERIPHERAL —	15-16 / 1600-6400	15-16 / 1600-6400	8-30 / 800-3200	8-30 / 800-3200	15-30 / 1600-3200
SUB'S COST ($/cm²) ———— PROTOTYPE ——	$300/cm² / $80	$10	$6-$10 / $2-$4	$6-$10 / $2-$4	$0.06/LAYER /cm²

Figure 10.2 MCM types and characteristics.

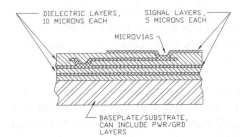

Figure 10.3 MCM-D thin-film substrate construction.

MCM-D (thin-film). By combining the power-handling ability of ceramics with the relatively low dielectric constant of polymers and the ultrasmall feature sizes of thin-film technology, MCM-D has become the category of choice for high-performance circuit applications. Conventional ceramic multilayer substrates serve as the power, and the ground and structural base on which a polymer dielectric multilayer buildup is constructed using relatively low-temperature conformal coating processes of spin or spray application. Vapor-phase disposition and lamination of partially cured polymer films are other techniques that can be used to build up the dielectric multilayer on top of the ceramic substrate (see Fig. 10.3).

Dielectric materials criteria used to select thin-film MCM-D are as listed:

1. The materials should be thermally stable to avoid uncontrolled outgassing during substrate construction at elevated temperatures.

2. They should have a very low moisture absorption.

3. They should be applied as a thin laminate with minimum surface planarity irregularities between layers.

4. They should be able to form adhesive bonds to a variety of substrate materials (silicon, alumina, aluminum nitride, glass-ceramic, or metal).

5. They should be metalized to adhere to the dielectric.

6. They should be chemically resistant.

7. Any reaction between dielectric and conductor material should be controllable.

Two material types are generally used to construct thin-film interconnections for MCMs: polymers and silicon dioxide. A variety of polyimide reformulation materials have been developed and tailored to meet specific polymer characteristics.

1. Fluorinated polyimide (fluoropolymer composite using randomly oriented glass microfiber and ceramic fillers) was formulated to reduce

the normal moisture absorption and lower the dielectric constant of conventional polyimides. These improvements were gained, however, at the cost of lowering solvent resistivity. Lower solvent resistivity results in difficulties during fabrication of the thin-film interconnections.

2. Polyimides with lower X-Y (axes) CTE have also been developed. These lower X-Y CTE formulations also result in lower water absorption and lower dielectric constant. A drawback with these formulations is poor self-adhesion and adhesion to other materials when surface adhesion modifiers are not used.

3. Improvements in planarization have been achieved by lowering the molecular weight of conventional polyimides; but lower mechanical properties, such as tensile strength and elongation, have been produced as a result. These preimidized polyimides, however, significantly improve surface planarity of the finished dielectric layer.

4. Photosensitive polyamide is another formulation that has been developed. This material was developed to reduce the number of fabrication process steps during the buildup of the multilayer interconnection portion of the substrate. Photosensitive polyimides, however, experience higher water absorption and reduced mechanical properties, including shrinkage during final cure. These polyimides are available only as laminates.

5. Other prominent polymers, being used or developed, are benzocyclobutene (BCB), polyphenylquinoxalin (PPQ), and polyquinoline. BCBs have a lower molecular weight with improved planarization and one of the lowest dielectric constants (2.7) of all the polymers, lower reaction with metals, and practically no water absorption during cure. On the down side, BCBs readily oxidize in air at elevated temperatures exceeding 150°C and lose thermal stability above 350°C.

6. PPQ polymer is also a low dielectric constant material (2.7) with improved resistance to fabrication chemicals.

7. Another member of the PPQ family of polymers being developed with lower dielectric constant (2.6), lower water absorption (0.15 wt %), and more tolerant manufacturability is polyquinoline.

Silicon dioxide (1) is chemically stable, (2) is physically strong, (3) has consistent material characteristics, and (4) adheres to most other thin-film materials. This material, however, has the following less desirable features: (1) a higher dielectric constant (4.6) than thin-film polymers, (2) requires insulation, (3) its as-fired surface is rough (16 μm) and needs polishing which is difficult to procure, and (4) requires chemical vapor deposition temperatures approaching 400°C.

There is a flurry of activity within the industry to develop dielectric materials that have all the desired characteristics ideally suited for MCM substrate application.

Other MCM materials. Alumina ceramic is a material with well-established processes that require relatively high capital investment and tooling costs. Resolution of conductors is limited by the rheology of thick-film ink and screen printing.

Aluminum nitride is less costly, has a number of suppliers, has a lower dielectric constant, and has a very high thermal conductivity (230 watts per meter per degree Kelvin), and laminate is cofired. It does, however, have inconsistent characteristics, dissolves in alkaline solutions, is not available in large sizes, and surface decomposition to aluminum oxide may not be suitable for nonhermetic packaging.

Silicones are less costly, easily used for high-density interconnections, extremely smooth and flat, and dimensionally stable. Silicones have higher thermal conductivity than alumina, and active and passive devices can be fabricated in the silicon substrate. Silicones are, however, fragile, and I/O fanout along edges needs to be limited; have increasing delay times; and require stress-relief gull-wing leads and environmental protection.

Conductors. Conductors used with MCM-C substrates usually consist of metals suitable for high-temperature cofiring processes, such as palladium, copper, silver, gold, tungsten, and molybdenum. MCM-D substrates use aluminum, copper, or gold. MCM-L substrate conductors are almost always copper.

Diamond films, the electronic wonder material. Diamond is a very hard material uniquely suited for electronics. It is the best solid material for thermal conduction (better than most metals) and, at the same time, one of the best electrical insulators. Diamond is optical and IR-transparent, has high electrical resistivity, is dopable (n and p type), has high electron mobility, and has a low dielectric constant.

Once fully developed as a film, it can be applied on the surface of substrates and used in intimate contact beneath components for thermal dissipation and as an insulation coat.

10.1.5   MCM interconnective alternatives

There are three alternative methods of terminating bare chips to MCM substrates: wire bonding, tape automated bonding (TAB), and flip-chip.

Wire bonding. Wire bonding, developed and proved by the hybrid industry, is the most prevalent chip interconnect attachment technique currently used in MCM. This technique is the most flexible and can terminate any, and all, chips as they are currently designed. Manufacturing wire-bonding equipment, suitable for high-production throughput, is readily available. Wire-bonded joints can be easily inspected, pull-tested, and reworked. This technique, however, does require extra area—two terminal joints per wire—and is a serialized manufacturing process.

Since bare chips cannot be tested until attached with leads, wire-bonded chips are not functionally verified until the MCM is tested. MCM yields, therefore, are determined by the combined yields of all the chips in the module. The higher the number of chips, the lower the MCM yield. For this reason the number of chips per MCM will be kept relatively low, thereby limiting the utilization of MCM technology.

Tape automated bonding (TAB). TAB is a series of identical, fine-pitch copper lead frames supplied on a continuous sprocketed, reel movie film tape made of polyimide with each movie frame containing one lead frame. The function of TAB is to terminate active chips to substrates and PWBs. Those ends of the lead frame that terminate on the chip are described as the inner-lead bonds (ILBs), and the other ends, that terminate on the substrate and PWB are known as the outer-lead bonds (OLBs) (see Fig. 10.4 for OLB termination).

Copper lead frames are repeated on a movie film frame basis. Tapes made from polyimide film, configured like movie film tape with sprocket holes along each edge, are bonded to copper foil and patterned and etched into individual lead frames repeated on a movie frame basis all along the tape's length.

Reels of TAB can be automatically terminated to chips on a frame-by-frame basis (see Fig. 10.5) and tested prior to attachment to substrates and PWBs. The ability to pretest and burn in chips prior to final assembly is a major advantage of TAB. TAB bonded MCM chips can be reworked.

The major problem with TAB is the lack of I/O standards on chips. Every chip, regardless of I/O quantities, has potentially different I/O positions, making standardization of TAB ILB pattern practically impossible. At $12,000 development costs per TAB pattern, nonstandardization of ILB will limit the extent of TAB usage.

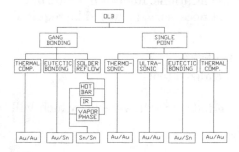

Figure 10.4 Outer-lead bonding options.

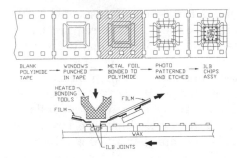

Figure 10.5 TAB fabrication and CMT assembly.

Flip-chip. Starting with a conventionally processed chip in wafer form, solder bumps are added by plating, or evaporation, of tin-lead on the I/O terminal pads on each chip of the wafer to prepare the chips for eventual attachment to substrates and PWBs. Most wafers not designed for flip-chips can still be bumped. Bumped chips are flipped and reflow-soldered directly to the substrates and PWBs. Flip-chips can be tested prior to reflow by pressing them into their assembly positions and testing the entire MCM. This approach, although adding extra steps, is an indirect way of pretesting chips, raising final MCM yields, and avoiding potentially destructive rework replacement.

Insertion of low-stress epoxy underfill between the chip and the substrate dissipates some of the stress and improves temperature cycling results.

Flip-chip technology could, in the near future, play a major role in electronics. This technology is already being used by large, vertically integrated companies who can get the chips bumped, while in the wafer stage, and made available to themselves. Flip-chip advantages are as follows:

1. Packaging density is as small as possible; zero area needed for chip interconnects.

2. Electronic performance is not affected by component-die interfacial lead length.

3. System will match existing chip I/O locations. Redesign of chip or establishment of I/O standards is not necessary, as would be required with TAB ILB.

4. Mass reflow soldering is relatively easy.

5. Repairability is possible with specialized tools.

This technology currently has the following drawbacks, which will limit its general usage:

1. Unavailability of bumped, pretested chips
2. Lack of standards defining bare chips
3. Assembly inspection possible only with use of X-ray laminography
4. Thermal transfer from chips to substrate
5. Fiducials needed on chips for placement accuracies

10.1.6   Optical MCMs

Electronics is moving to increasingly high frequencies. Communicating signals at higher and higher frequencies from device to device, module to module, and workstation to workstation through metal conductors will soon become impossible. Metal conduction is already inhibiting machine function at 100-MHz frequency. Optical transmission, on the other hand, is just getting started at 1-GHz frequency.

Local area networks (LANs) that had bandwidths to service a multiple array of workstations, at last year's workstation performance levels, can support only a couple of workstations at today's performance levels. Optically coupled LAN, in place of the current metal-coupled LAN, would immediately solve the problem of limited bandwidth.

Devices that convert electrical signals to optical signals and optical signals to electrical signals will soon become the components of the 1990s as optically based LAN systems come on-line. These new components will likely be MCMs with optical and electrical I/Os and, in time, include receiving and transmission logic. This is, perhaps, the largest market ahead for MCM.

10.1.7   MCM industry infrastructure

MCM industry infrastructure, in the author's opinion, is still in the process of building and has a considerable way to go before MCM can become a widely used technology. The vendor support base in the following areas needs to be either established or greatly improved:

1. Availability of tested bare chips
2. Low-cost substrates
3. Low-cost module packages
4. Rework and repair procedures and tools
5. High-volume manufacturing machines and tools
6. Specific CAD software tools
7. Standardized test techniques
8. Standardized test software

9. Standardized I/O on active chips

10. Time-to-market for completed MCM suppliers

Vertically integrated companies, such as IBM, DEC, Texas Instruments, and Motorola, who can build in an infrastructure within their own company, have a large advantage over their other, non–vertically integrated, competitors. The only option open to those other competitors, at this time, is to create their own infrastructure by forming strategic alliances or partnerships in specific areas with other companies as needed.

10.2   Chip-Mounted Technology

Chip-mounted technology (CMT), previously known as "chip-on-board" (COB), is a relatively new SMT subtechnology dealing with the design, attachment, and assembly of unpackaged active chips.

Unpackaged chips, although very brittle and fragile, can mechanically survive without a package. However, nonsealed or nonencapsulated chips cannot survive. In the presence of moisture and elevated temperatures, unsealed chips can catastrophically fail within minutes.

Two methods are used to seal chips: (1) encapsulation at assembly and (2) presealing.

10.2.1   Encapsulation at assembly

After chips are attached, usually with silver-filler epoxy for good thermal and electrical conductivity, ultrasonically wire-bonded with aluminum wire (PWB copper pads are normally nickel-plated with a gold flash overcoat), pull-tested to 10 to 15 g, and electrically tested, the chips and immediate surrounding area are encapsulated with low ionic epoxy and cured for 12 h (see Fig. 10.6). After curing of the encapsulation, the assembly proceeds through the normal SMT assembly line for the mounting and soldering of all other components.

Encapsulation at assembly has two problems. First, the chips are untested at the time of attachment and, except for destructive removal, nonrepairable after assembly. Reliability of bare, untested chips is relatively poor. The probability of one or more chips not operating is un-

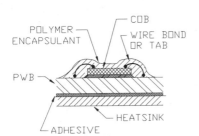

Figure 10.6 CMT assembly encapsulation.

acceptably high for complex circuits. It can also be unacceptable for simple circuits using chips that have historically poor wafer yields. The second problem is the lack of long-term moisture protection of epoxies. Encapsulation at assembly is being used for non–environmentally challenged products and those with a lower life-cycle expectancy.

10.2.2 Presealed chips

Chip interconnection pads are bumped, with gold bumps or tin-lead bumps, and then connected to the inner leads of a TAB frame, tested, and sealed with a protective coat(s). In this arrangement, the sealed, TAB-bonded chip is at the same status as a tested, packaged chip; and except for the critical handling sensitivity of sealed chips and the closer lead pitch of TAB, they are interchangeable.

Sealed chips are gaining popularity as the sealing, often multilayered of various material types, improves. Developmental sealing layers, topped with proprietary gel coats, have already passed 10-year accelerated life testing in accordance with MIL-STD-883. The telecommunication and automotive segments of the industry are about to commit more extensive usage of sealed chips into their product designs. As these products gain "field" exposure, and the coatings are made commercially available, sealed chips will experience widespread industry acceptance, including the military.

Except for interconnecting pads, the entire chip is normally coated with a moisture barrier passivation layer. Passivation layers are vulnerable to pinhole voids and consequently cannot be entirely trusted to protect the chip from moisture. Moisture penetration at the nonpassivated interconnection pads can be prevented by the application of the interconnection bumps (see Fig. 10.7).

Moisture. Moisture per se is not the problem. Chips can continuously function submerged in a bucket of deionized water. It is ionized water or moisture that causes the problem. The ideal coating would be one that would filter and deionize the water prior to its full penetration down to the chip surface.

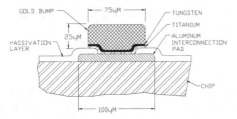

Figure 10.7 Chip interconnect bump.

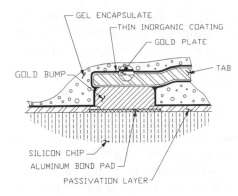

Figure 10.8 Four-layer chip seal coats.

Silicon gel coat. Silicon gel coats, having a polarized molecular structure in which all the molecules attach themselves to the chip surface in the same polarity, forming a barrier to water molecules, have been developed by several companies and are now, or could soon be, commercially available on the open market (see Fig. 10.8).

Even though gel coats form a molecular barrier to water, it is assumed that gel barriers do not properly form over pinholes in the passivation layer and consequently present a "chimney" for water to enter the chip. For this reason undercoats are applied before gel application. Diamond films and other inorganic materials are gel coat substitution candidates. Efforts are also under way to improve the existing passivation process to make it pinhole-free.

Gel coats outgas and therefore must be used with some caution, especially in outer-space applications.

TAB standardization. TAB will not reach its full potential until chip suppliers establish standards governing interconnection locations to the same degree as package standards have been. At present, there could be 10 different TAB patterns for 10 different 20-pin devices. With TAB lead-frame pattern development costing $12,000 each, a typical assembly with 50 different active devices may not be economically feasible for lower-production quantities. A user's warehouse could soon be filled with partially used TAB reels.

Miniaturization of PWB features. CMT will not reach its full subminiaturization and circuit speed potential until multilayer substrates (or PWBs) can be economically mass-produced with 75-μm conductor width and spacing.

Manufacturing impact. Dealing with fragile CMT in mass production will be a major change for manufacturing facilities and personnel skill mix. Rework and repair will also be a new learning experience.

Once CMT is fully developed and implementable, it will quickly become the technology of choice for all segments of the industry. It will replace most packaged chips and MCMs.

10.3 Solder Substitute

Solder is a terrible *material*, as seen from a mechanical engineer's point of view. Soldering is a terrible *process*, as seen from a manufacturing engineer's point of view. Solder is structurally poor, fatigues very quickly, is a metallurgical quagmire when encountered by precious metals (wicking can make even moderate control difficult), adheres to very few materials (and then only to superclean oxide-free surfaces), contains EPA (U.S. Environmental Protection Agency)-condemned material (lead), and has a relatively low melting temperature. Soldering operation requires 230°C, needs a generous amount of freely applied acid-based etchant, requires very careful control of time and temperature, requires an inordinate amount of surface preparation, and is costly and demanding. Furthermore, certified operators are required, the process requires ventilation, necessary machine maintenance is relatively high and expensive, and cleanup is difficult and getting worse with new EPA regulations. The selection of components and materials is dominated by the high temperatures and thermal shock survival requirements imposed by soldering processes. Soldering processes bring a complexity of conflicting materials to the manufacturing arena that is unlike any other process. Acids, bases, salts, metals, alloys, electrical current, high temperatures, toxic fumes, flammables, solvents, and compound mixtures all come together at some point in support of the soldering process. Industry says, "there has got to be a better way of achieving mass termination bonding of electronic interconnections." In response, the industry has turned to, among other things, conductive adhesives.

Epoxy conductive adhesives, having successfully been used as a solder substitute in the hybrid industry, are now being pursued as a possible candidate for mass soldering of SMT assemblies. Epoxy adhesives can be cured at either low or high temperatures. Low-temperature curing would resolve the problem with heat-sensitive components but would require longer cure time. High-temperature curing could use the same IR soldering machines and profiles as the normal soldering production throughput. This approach would, however, continue the high-temperature exposure for the components and materials. Epoxies cured at solder reflow temperatures require less than 10 min, and those cured at 125°C require four to five times that amount.

SMT epoxy conductive adhesive joints perform better than solder joints in fatigue life and thermal shock when the epoxy formulation is based on high-strength, structural adhesives, rather than the conventional hybrid formulation. Rheology of epoxy conductive adhesives, with its inherently low surface tension (one-fifth solder) and smaller conductive particle size, is ideally suited for stencil, screen, or syringe application.

Unlike solder, epoxy adheres to most materials with far less surface preparation than do those required by solder. But like solder, silver, the conventual conductive filler, migrates from the adhesive to adjacent conductors in the presence of moisture and a sufficient voltage differential.

SMT assemblies, with epoxy adhesive conductive joints, can be repaired by simply applying 150°C heat and lifting the faulty component. Replacement joints can be formed without removing adhesive residue.

With the proper development, performance tests, and process tests, conductive adhesives could become the mass attachment technique that replaces solder.

10.4 Heat Sinks

Advance packaging often involves thermal management that requires lightweight heat-sink materials. Composites are becoming the material of choice for avionic thermal management. See Fig. 10.9 for SEM-E-size heat sinks.

10.5 Updating Engineers

For engineering and manufacturing, keeping abreast and keeping competitive has increased at a faster pace. Keeping abreast is difficult but necessary. Up to now, the responsibility for keeping individuals abreast has mostly been the responsibility of the individuals themselves. Companies have kept somewhat abreast by targeting specific areas they label as their future "zones of excellence." But these zones of excellence are segmented and specialized and ignore the continuum

TECHNOLOGY	HEATSINK MATERIAL	POWER (ALUM=1)	COST	WEIGHT (g)
SMT	CIC ―――	0.87	$325	1180
	ALUMINUM ―	1.00	$300	717
MCM	EMBEDDED HEAT PIPES	1.5	$350	699
	BERYLLIUM COMPOSITE	1.11	$800	590
	GRAPHITE/ EPOXY	0.97	$1000	577
	GRAPHITE/ ALUMINUM $_1$	1.07	$1000	636
MCM/COB	GRAPHITE/ ALUMINUM $_2$	1.25	$1400	636
	AIR COOLED HOLLOW CORE	1.43	$400	522
	LIQUID COOLED HOLLOW CORE	3.40	$400	563

Figure 10.9 Heat-sink power, cost, and weight—SEM-E size.

necessary with all designs. Keeping up with the broader aspect of design is necessary for the design engineer and the manufacturing engineer who are charged with system designs and system fabrication and is necessary for the company employing them.

It will become increasingly necessary for companies to formalize the process of helping individual engineers keep relevant. Changes within the state of the art are moving at such an increased pace that individual engineers could soon not have the resources or spare time to keep up. Company resources may be necessary to support an engineer's efforts if the company is to remain competitive. Companies who see the need to help could consider incorporating one or more of the following:

1. Bring into the company recognized industry experts to teach and upgrade the staff in specific areas.

2. Use video training materials from industrial societies and/or suppliers.

3. Assign specific individuals to serve as members of technical societies and as sentinels and disseminators of SOA changes within the industry.

4. Contract university professor(s) to survey industry growth and upgrade staff.

5. Ask the company R&D principal engineer to share his or her experience and knowledge of special projects.

6. Assign individuals to research and/or survey specific areas of immediate concern and prepare a dissertation document on the subject.

7. Recognize and award individuals for extra effort of keeping abreast.

8. Keep a "lessons learned" journal updated and available.

10.6 Single-Pass Reflow of IMT and SMT Type II Assembly

Cost-saving reduction of process steps is stimulating efforts to eliminate the costly wave soldering from Type II assemblies. A single-pass reflow process is being developed to simultaneously mass-solder IMT and SMT components. Getting enough solder paste into the plated through-hole without having the IMT component leads dislodge the paste during the insertion process and ending with an accepted reflowed IMT-type solder joint is the challenge. Pretinned leads and pads, automatically placed preforms, automatic syringe dispensing, and stenciling or squeegeeing techniques are some of the methods being reviewed and developed.

Changing IMT solder joint standards is also being reviewed. Nonprotruding leads that penetrate only two-thirds of the hole depth (eliminating paste push-out) and small solder fillets are offered as possible standards. A nonclinched lead is in keeping with SMT joints where the solder serves the mechanical and electrical functions. Heavier IMT components can be automatically adhesive-bonded when the properties of the solder joint need to be mechanically supplemented.

The added thermal mass of the larger IMT component bodies and leads will impact the reflow system; it may even be beyond the thermal capacity of most systems.

10.7 SMT by the Year 2000

SMT packaging techniques, as currently practiced, will be significantly altered by new materials, new processes, and new subtechnologies. Following are some of the factors that will affect future packaging:

1. *3D packaging.* Three-dimensional packaging will also become a viable packaging alternative. Flip-chip technology and total-plane interconnects will permit 3D stacking of chips and interplane wiring.

2. *Solder substitute.* Molten solder will begin to be phased out and replaced by low-temperature organic conductive adhesives. Precision lasers will be commonplace on the assembly line for very-fine-pitched joints, rework and repair, assembly-line measurements, and automatic lamination alignment tools.

3. *Fiber optics.* Fiber optics will replace many wire conductors, even within localized systems and assemblies. Optical receivers and transmitters will be prominent components on PWBs. Speed, elimination of crosstalk and propagation delays, security, and signal capacity will all force the use of optics.

4. *Fiber-optics terminals.* Relatively small, specialized metal packages, with superaccurate cavities and tailored CTE, will be grown to precision dimensions for fiber-optic terminations and splitting and also for millimeter wave devices.

5. *Diamond film.* Diamond film will become a very important material used in optics, sensors, thermal management, and, because of its dopability, perhaps in circuit logic elements. Superconductors will also play a key role, but in a few specialized areas involving magnetics and superhigh speeds.

6. *Future factories.* Electronics factories will be more like present-day hybrid facilities than present-day PWB assembly lines. Workforce skills will be more demanding and specialized. Factories will be smaller but will cost more.

Author's private view of MCM and CMT future. There are professional people in electronics marketing research whose business is to predict the industry's future. I, the author, am an amateur in the field of industrial predictions. I do, however, have a private opinion, contrary to the professional opinions, based on my own experiences as to the fruition of today's trends by the year 2000.

I believe that CMT on PWBs will dominate the industry by 2000 and that MCM usage will thereby be diminished. There are three factors holding back CMT from being the dominant technology today.

1. *Ten-years-plus hermetic seal of bare chips not yet available.* Two approaches are being pursued by industry to achieve the 10-year hermetic seal goal. Silicon gel coatings, and others, are in the last stages of full development. The other approach, now being researched, is to develop pinhole-free passivation. Either one or both of these solutions promises to be validated by middecade.

2. *Available pretested, TAB-bonded, sealed chips.* There is no technical reason why CMT components can't be made available once the seal coat, or pinhole-free passivation, is qualified. Neither is there a business reason for holding back CMT components, except for the reluctance of the suppliers to standardize the chip interfaces to match standardized ILB and OLB of TAB. Worldwide competition will soon resolve the supplier's reluctance.

3. *Large PWBs with 75-μm line and space.* Small PWBs with 75-μm line and space are being produced for MCM today. The problem with building large boards in production quantities is a combination of material and process difficulties. PWB laminate materials with improved stability are being developed and could soon be available. Direct imaging is likely to be perfected very soon. Copper foil at 4 to 5 μm thick is already obtainable. With semiadditive processing-controlled lithography and laser-drilled blind vias in solder pads, large boards with subminiature features can be economically ready before the mid-1990s.

I believe that once CMT components are readily available with matching PWBs, the need for most of today's MCMs will cease to exist. The cost-effectiveness and performance effectiveness of CMT will simply eliminate all but the very-high-density, very-high-speed MCM-D devices (see Fig. 10.10). The industry's prediction differs (see Fig. 10.11).

Market forces and customer expectations appear to be far more demanding in this decade than in earlier decades. These demands will force competition to cause an accelerated race toward ultrasubminiaturization and more powerful functional performance.

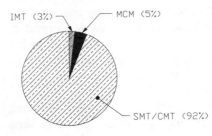

Figure 10.10 Packaging by year 2000—author's prediction.

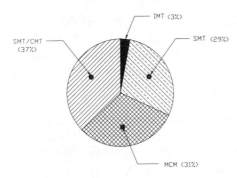

Figure 10.11 Packaging by year 2000—industry prediction.

Bibliography

Adema, Gretchen M., Michele J. Berry, and Iwona Turlik: "Dielectric Materials for Use in Thin Film Multichip Modules," *Electronic Packaging and Production,* February 1992, pp. 72–76.

Bolger, Justin C., John M. Sylva, and James F. McGovern: "Conductive Epoxy Adhesives to Replace Solder," *Surface Mount Technology,* February 1992, pp. 66–70.

Burkhart, Art: "Recent Developments in Flip Chip Technology," *Surface Mount Technology,* July 1991, pp. 41–44.

Costlow, Terry: "MCM Shakeout Continues," *Electronic Engineering Times,* April 6, 1992.

Editorial: "Positive Outlook for the MCM Industry," *Circuit Assembly,* September 1991, p. 10.

Editorial: "Distributor Eases into Bare Die Supply for MCM's," *Military and Aerospace Electronics,* September 1991.

Ginsbery, Gerald: "Multichip Modules Gather ICs into a Small Area," *Electronic Packaging and Production,* October 1988, pp. 48–49.

Hodson, Timothy L.: "Using Existing Technologies to Process Multilayer MCMs," *Electronic Packaging and Production,* August 1991, pp. 48–50.

———: "Bonding Alternatives for Multichip Modules," *Electronic Packaging and Production,* April 1992, pp. 38–42.

Hong, Stephen, and Jason Pei: "Cooling MCMs with Pin Fin Heat Sinks," *Electronic Packaging and Production.*

Kozak, Al: "MCMs in Telecommunication Designs," *Surface Mount Technology,* March 1991, pp. 46–49.

Leilowitz, Joseph D. "A Novel Concept for a Multichip Module," *Surface Mount Technology,* May 1992, pp. 45–49.

Lezotte, John: "Optical MCMs Lead the Way," *Electronic Engineering Times,* June 10, 1991, pp. 68–69.

Ng, Dr. Lee H.: "MCM Cost: The Volume-Yield Relationship," *Surface Mount Technology,* March 1991, pp. 56–61.

Nielsen, Arne: "Chip on Board Technology," *PC FAB,* October 1988, pp. 80–84.

Perry, Dave: "Advanced System Technologies: Tailor-Made for High Performance," *Surface Mount Technology,* September 1991, pp. 22–24.

Sabatini, Jim E.: "MCM Design—Using High-Density Packaging," *Surface Mount Technology,* March 1992, pp. 18–19.

Tuck, John: "MCMs: A Year Later," *Circuit Assembly,* March 1992, pp. 24–25.

Glossary

accelerated aging A test that elevates parameters such as voltage, temperature, humidity, or chemical exposure above normal operating values to generate deterioration aging data in a relatively short time.

additive process Deposition of conductive materials on unclad base material in desired locations as opposed to the subtractive process where clad conductive material is etched away from nondesired locations.

air gap The space between the bottom surface of SMT components and the top surface of substrates.

air knife High-pressure airflow directed across molten soldered surfaces, expelling excess solder and generally leveling the surface of the remaining solder.

algorithm A set of well-defined programming rules that solve a problem in a finite number of steps.

alumina Aluminum oxide ceramic material used to construct electronic substrate, chip component bodies, and component packages.

ANSI American National Standards Institute, 1430 Broadway, New York, NY 10018, USA.

aramid A generic description of high-temperature aromatic polyamides.

artwork A 1oft drawing or photographic tool of PWB conductor routing.

ASIC Application-specific integrated circuit; a custom-made VLSI.

aspect ratio Diameter-to-length ratio of PTHs in PWBs.

ASTM American Society of Testing and Materials, 1916 Race Street, Philadelphia, PA 19103, USA.

ATE Automatic test equipment.

automatic placement Placement of SMT components on PWBs into preapplied solder paste or adhesive with the use of automated machinery.

azeotropic An adjective describing a compound which retains its liquid-state composition when in the vapor state.

ball bond Initial bond of thermocompression wire bonding operation between chips and substrates.

bed of nails   A matrixed array of spring-loaded test probes fixed to a planar base.

beryllia   Beryllium oxide ceramic material used for its favorable thermal conductivity characteristics as a substrate.

beryllium copper   An alloy of beryllium and copper used in lead frames for its favorable heat dissipation.

BIT   Built-in-test; features designed into products to enable testing at higher assembly levels.

blind vias   Partial vias that interconnect the outer surface(s) substrate conductors with internal conductors.

bond   The area of molecular fixing of two materials that provides the electrical interface between them.

bonding pad   The metalized site at which a bond is made.

bridge   A short circuit usually caused by inadvertently applied solder between two circuit paths.

bump   An ultraminiaturized, hemispherically shaped, metal deposit, usually tin-lead alloy or gold, applied to chip or TAB (tape automatic bonding) bond pads to enhance interconnection.

buried vias   Partial vias that interconnect two or more internal conductor layers without extending to either outside surface.

burn-in   An electrical stress test to screen out components with infant mortality latent defects.

butt lead   A SMT component lead configuration that radiates outward and downward and abruptly ends with a cutoff just below the bottom surface of the component.

CAD   Computer-aided design; in SMT, refers to the design process in which much of the conductor routing is automatically done by a computer based on a circuit network input.

CAE   Computer-aided engineering; in SMT, refers to worst-case circuit and thermal analysis performed by the computer.

CAM   Computer-aided manufacturing; in SMT, refers to computer-based manufacturing workload analysis, planning, etc.

castellation   Metalized, vertical flute along the outer edge of a leadless chip carrier serving as the component I/O interconnection.

CCA   Circuit card assembly (alternate name for PWB assembly).

CCC   Ceramic chip carrier; a generic term for a LDCC or a LLCC component.

centipoise   A unit of measure of the coefficient of dynamic viscosity.

ceramic   A hard, brittle material used to construct component bodies and assembly substrates, made by firing clays or other like materials.

CFC   Chlorofluorocarbon; very effective nonpolar postsoldering cleaning solvents, allegedly the cause of reduction in the global ozone layer, to be phased out of production operations by the Montreal Protocol by mid-decade.

characteristic impedance   Ratio of voltage to current as a resistance to signal flow.

chip   A small, thin, rectangular die of pure silicon on which active and passive circuit elements have been microelectronically deposited and interconnected into a functional circuit. Also a term used to describe a small, rectangularly shaped ceramic passive component.

chip carrier   An active SMT component package, either square or rectangular, with leads or terminations on all four sides.

CIC   Copper-invar-copper; a coined laminate of three metal layers whose composite CTE is tailored to constrain PWBs, thereby limiting solder joint stress. It is also used in a second role for heat dissipation.

CIM   Computer integrated manufacturing; the top computer system that electronically ties all the individual computer subsystems into a companywide central network of free-flowing data for the purpose of monitoring and controlling production.

circuit element   A part of an IC which contributes directly to its electrical characteristics (resistors, capacitors, thermistors, etc.).

circuit gate   A circuit subdivision having multiple inputs and a single output triggered by a prescribed combination of input signals.

CMC   Copper-molybdenum-copper; a slightly improved alternate for CIC.

CMT   Chip-mounted technology; the revised term for COB.

COB   Chip-on-board; a passé or outmoded term referring to unpackaged chips mounted directly on PWBs.

component   Passive, active, or other types of discrete electronic devices.

component density   The quantity of components on a PWB per unit area.

condensation soldering   See **vapor-phase soldering.**

convection soldering   Oven soldering method in which hot air or hot gas is the chief heating medium.

contact printing   Screen or stencil printing performed with the screen or stencil mask resting directly on the substrate without the benefit of snap-off action.

core   A layer (or layers) of material laminated in the center of PWB primarily as a mechanical element to constrain the board CTE or dissipate heat.

CQFP   Ceramic quad flat pack; a fine-pitch, hermetically sealed package.

creep   The dimensional change of a material over time while under a mechanical force.

crosstalk   An undesirable electrical coupling induced from one circuit node to another.

CTE   Coefficient of thermal expansion; the amount of expansion and contraction individual materials experience when subjected to temperature changes.

deionized water   Water devoid of ions.

dewetted   A surface condition describing an area in which molten solder covered but then withdrew as a result of inadequate wetting conditions on the surface.

die   Semiconductor circuit element fabricated by batch diffusion process on a silicon wafer which is then "diced" (cut) into individual (die) components. A term used interchangeably with **chip.**

DIP   Dual in-line package; an insertion-mounted active component type with lead along two opposite sides.

DMA   Dynamic mechanical analysis; a quantitative test method that measures the change of stiffness of resin laminate with change in temperature.

DoD   Department of Defense; the agency directing the efforts of the U.S. military.

dopable   Capable of being used as a semiconductor circuit element by impregnating with a doping agent. Common doping agents for germanium and silicon are aluminum, antimony, arsenic, gallium, and indium.

DSC   Differential scanning calorimetry; a quantitative test that determines the glass transition temperature of resin materials.

ECA   Electronic circuit assembly; a module consisting of two CCAs mounted back to back as a single unit.

EIA   Electronics Industries Association, 2001 Eye Street, Washington, DC 20006, USA.

EIAJ   Electronics Industries Association of Japan, 2-2, Marunouchi 3-Chome, Chiyoda-Ku, Tokyo 100, Japan.

eutectic   A term describing the unique metallurgical phase of tin-lead solder alloy in which the alloy, at the eutectic temperature, switches directly from solid to liquid without going through a transitional plastic phase.

fillet   See **solder fillet.**

flat pack   A relatively thin SMT active component package style with unformed and untrimmed leads that at assembly are formed into the gull-wing configuration.

flip-chips   Chips with bumps added.

footprint   A pattern on PWB formed by a group of individual solder pads arrayed to match a particular SMT component type.

Ga As   Gallium arsenide; a semiconductor technology material used in lieu of silicon or germanium for very-high-speed circuit chips base.

gull-wing   A SMT component lead configuration radiating straight outward from the outer edge of a component having a pronounced vertical offset.

IC   Integrated circuit; a microelectronic chip composed of semiconductor devices that form a single or multiple electronic circuit function(s). The term is often used to describe the packaged device containing the IC chip.

IEC   International Electrotechnical Commission, P.O. Box 131, 1211 Geneva 20, Switzerland.

IEEE   Institute of Electrical and Electronics Engineering, 345 East 47th Street, New York, NY 10017, USA.

IEPS   International Electronics Packaging Society, 114 North Hale Street, Wheaton, IL 60187, USA.

IIMT   International Institute of Manufacturing Technology, Box 549, MCV Station, Richmond, VA 23296-001, USA.

ILB   Inner-lead bond; the TAB interface that interconnects chips.

impedance match   Impedance of a component or circuit, equal to the internal impedance of the source or the impedance of a transmission line during operation.

IMT   Insertion-mounted technology; an acronym that embodies the electronic packaging method dealing with components that are inserted and terminated with the use of the PWB Z axis.

infrared   Electromagnetic radiated energy band used in SMT for soldering and thermal measurement sensors.

intermetallic   An undesirable metallurgic amalgamation formed at the interface surface of two or more molten metals.

I/O   Input/output; term used to describe the electronic functional interface of a device or assembly. This term is used interchangeably in the electrical and mechanical sense.

IPC   Institute for Interconnecting and Packaging Electronic Circuits, 7380 North Lincoln Avenue, Lincolnwood, IL 60646, USA.

IR   Acronym for infrared.

ISHM   International Society for Hybrid Microelectronics, P.O. Box 2698, Reston, VA 22090, USA.

JEDEC   Joint Electron Device Engineering Council, arm of EIA.

J lead   A SMT component lead configuration that radiates outward and downward with the end curled beneath the component body.

junction   The boundary between two different regions (n and p) of a semiconductor device permitting current to flow in one predominant direction.

LAN   Local area network is a network of wires and cables interconnecting interdependent machines and computers within a given factory area(s) or site(s) for purposes of automatic data transfer within the network.

LSI   Large-scale integration; the combining of many individual IC functions into a single, larger function on one die.

Matériel   An encompassing term covering raw material, parts, components, tools, and expendables.

MELF   Metal electrical face; refers to a cylindrically shaped SMT component with metalized end caps serving as the I/O interconnections.

mesh size   A measurement of the open spaces within a screen in terms of quantity of openings per linear length.

migration   The transfer of ions from one material into an adjacent material having less ions. A condition in which ionic contamination left on the surface of PWBs can produce dendritic growth, leaching, etc.

MOS   Metal oxide semiconductor; a device that uses an oxide of the substrate material as the insulating layer. For silicon substrates the insulator is silicon dioxide (CMOS—complementary MOS; VMOS—vertical MOS).

NASA   National Aeronautics and Space Administration, P.O. Box 8757, Baltimore, MD 21240-0757, USA (for technical data).

OEM   Original equipment manufacturer.

OLB   Outer-lead bond; the TAB interface that interconnects the substrate.

outgassing   Emission of entrapped air or gases from solder paste, and other materials, produced by elevation of temperature or vacuum.

oxidation   The addition of oxygen on the surface of a metal in which metallic ions are formed.

passivation   Growth of an oxide layer deposited on a semiconductor device to protect it from the environment.

PLCC   Plastic leaded chip carrier; an active SMT component in a plastic package with leads radiating from all four sides.

polar solvents   Solvents that are ionized to the extent of being electrically conductive and hence capable of dissolving polar compounds.

polyimide   High-temperature thermoplastic resins used for PWBs, wire covers, and other purposes.

ppm   Parts per million; a unit used to indicate the quantity of unacceptable or distinguishable parts in a grouping of one million parts.

PQFP   Plastic quad flat pack; an active SMT component in a plastic package designed to enhance mechanized placement with leads radiating from all four sides.

PTH   Plated through-hole; the structural mounting hole and electrical interface on PWBs for insertion-mounted electronic components. It can also be the through-board electrical interconnection for SMT PWBs.

PWB   Printed wiring board; the substrate structural and laminated wiring plan on which electronic components are assembled.

reflow soldering   The process of thermally raising solder suspended in paste or solder preforms to the molten stage and forming electronic interconnections.

resin High-molecular-weight organic material.

semiconductor A material that is neither a classical conductor nor an insulator but whose electrical characteristics are somewhere in between the two.

sequential lamination Lamination process in which previously laminated and via hole–plated groups of layers are combined into a single laminate and hole-plated.

sintering The process of bonding metal or other powdered materials by pressing and heating to form a cohesive device.

slumping Relaxation of solder paste prior to reflow, resulting in loss of desired paste profile definition.

SMT Surface mount technology; a term that embodies the electronic packaging method dealing with components that are mounted and terminated without the use of the PWB Z axis.

SO Small outline; a term used to describe small, active, SMT discrete components that have leads along two opposing sides.

substrate In SMT, the planar structure that serves as the mounting plane for electronic components and can include interconnection conductors.

swim In SMT, the inadvertent movement of components away from their intended alignment caused by dynamic forces during the soldering process that renders the component buoyant and hence prone to buffeting.

T_g A symbol for the glass transition temperature of materials at which point the material's CTE switches from a relatively linear reaction with temperature rise to a rapidly increasing, nonlinear reaction.

TGA Thermogravimetric analysis; a quantitative test to determine rosin cure status by raising the temperature and observing weight variations.

thermal management Designed-in packaging features that control thermal dissipation.

thermocompression bonding A bonding process using the combination of heat and pressure to form a "blacksmith" weld nugget between two materials such as gold or aluminum wire to a chip or substrate.

thick film Electronic elements formed on a substrate by screen or stencil-printing thick film (>100 μm) of conductive, resistive, and dielectric pastes and/or inks and then fired at elevated temperatures to drive off binders and sinter the solids.

thin film Electronic elements formed on a substrate by vapor deposition or sputtering thin film (<5 μm) of conductive, resistive, and dielectric compounds.

thixotropic A characteristic of a compound that becomes liquid when stirred or agitated but sets again to its original consistency when the agitation is stopped.

tombstone The orientation of a passive chip component where one end is terminated with solder and the other end is unterminated and tilted above the surface.

trace   In SMT, a PWB conductor.

ultrasonic bonding   A bonding process using the combination of pressure and ultrasonically induced scrubbing action to form a molecular bond.

ultraviolet cure   Adhesive or ink curing process that accelerates molecular cross-linking by exposure to ultraviolet radiation.

via   Metalized holes, partially through or all the way through the substrate, that serve as the electronic interface between substrate conductor layers.

VLSI   Very-large-scale integration; large-scale integrated circuits containing 100 gates or more.

voids   Spherical cavities in solder joints, normally formed by outgassing of entrapped gases or moisture in solder or in solder paste.

wafer   A thin slice of pure silicon, or others, on which 1000 or more individual circuit functions are fabricated by batch diffusion processes.

waffle packs   A matrix tray of pockets used to package delicate SMT components for shipping, handling, and assembly processing.

warp   In SMT, a weaving term applied to the thread that runs lengthwise with the fabric.

wave soldering   Application of solder in the molten phase that forms the electronic interface between the component and PWBs as well as delivering the thermal energy needed to perform the soldering.

wedge bonding   A thermocompression or ultrasonic bond made with a wedge-shaped tool, usually the second bond made on the wire interconnecting chips to substrates.

wetting agent   A compound used to reduce the surface tension of a liquid and improve its function.

wicking   Rising of molten solder up a terminal or lead as a result of capillary forces.

woof   A weaving term applied to the thread that runs perpendicular to the warp threads in a fabric ("weft" and "pick" are alternate terms).

Z axis   The axis perpendicular to the plane (X-Y axes) defined by the mounting surface(s) of SMT assemblies.

A

Total Quality Management (TQM) Program

To meet the competition in contemporary markets, quality must dominate. The first and highest priority in all aspects of company endeavor should be to produce and deliver quality, defect-free products and service to the customer on schedule. Each function and stage in the company manufacturing and service must work together with suppliers to both receive defect-free matériel and produce defect-free products.

Although the following presentation is made as if to a large, OEM (original equipment manufacturer) company, it can be scaled down to meet the needs of much smaller companies.

Management commitment. Top-level management should initiate the preparation of a TQM strategic operating plan for the company. To formulate such a plan, a steering committee of functional directors from engineering, manufacturing, quality, and procurement and a senior member of top-level management as chairperson with the quality director as cochairperson, would be formed and empowered. This steering committee would meet, independent of other meetings, to deliberate and formulate the company's position and policy on TQM and perform the following functions:

1. Prepare a strategic operating plan with short- and long-term quality goals with functional responsibilities and timelines included.
2. Determine the methods of overall company measurements and define each goal accordingly.

3. Prepare and deliver the strategic operating plan for approval by top management and possibly the board of directors.
4. Disseminate the goals and measurement methods to functional managers.
5. Oversee the company's implementation of TQM and monitor its progress.
6. Prepare periodic reports to top management.
7. Form and empower an implementation working group with a representation from each functional director. The functions of this group are to
   a. Serve as the coordinating center for all TQM initiatives
   b. Determine the means, requirement, and staffing necessary to accomplish the goals
   c. Serve as the focus for all required cross-functional TQM activities
   d. Prepare detailed plans and schedules to complete the implementation
   e. Assign necessary working groups to accomplish the desired results
   f. Report to the steering committee progress attained and actions to be taken

Functional directors will be responsible for all TQM functions within their own organizations and functional procedures adjusted as necessary to accommodate TQM.

Customer involvement.   Customers should be made privy to the TQM plan, status, and results, especially those customers with whom there is a strategic alliance or contractual obligation.

Audits and reviews.   The audit process is one of the primary means that customers use to evaluate a supplier's ability to perform. Audits and reviews are becoming a way of life up and down the supply chain. Recognizing this situation, senior talented management teams are appointed to conduct internal and supplier audits and reviews and to receive customer auditors and act as the point of contact for all customer audits within the company. Team appointments are based on skill, background and experience, and poise with the customer. Each team member is responsible for certain areas of expertise and ensuring that documents are correct and available to the auditors and that access to requested areas is provided.

Cost of poor quality.   Determining the cost of poor quality is enlightening and should be done to serve as a baseline for all future quality improvements. Poor-quality costs need to be separated from total quality cost.

Total quality costs are subdivided into the following three categories:

1. *Prevention.* This category includes all the activities associated with the prevention of defects in deliverable products or services. The cost of statistical process control, to be discussed in detail later, is an example of the cost in this subgroup.

2. *Appraisal.* Costs associated with measuring, evaluating, testing, or auditing products or services to assure conformance with quality standards, and performance requirements are included in this category.

3. *Failure.* This category includes the costs required to evaluate, correct, or replace products or services that do not conform to requirements or customer needs. Items that fit this category include rework, scrap, repair, parts inventories, warranties, and—the largest cost of all—lost customers and the associated effort to reestablish them.

Data-based information and management system. Development and implementation of an integrated total product and service data-based information and management system is a key element of a TQM system and essential to achieving a quantum upgrade of engineering, manufacturing, and support capabilities. A companywide data-based system that enables capture and access of all data related to design, production, and field support is necessary for rapid deployment of TQM and quick turnaround market response of products or services. The larger the company, the more urgent is the implementation of the data-based system. This system will support electronic data interchange between functional areas within the company and suppliers and vendors outside the company and could be tied to key customers.

A data-based information and management system as described here is the key element that enables a comprehensive TQM system to function smoothly and, with time and attention, flawlessly. With such a system, all the measurement and data management capabilities needed to monitor and control, in real time, design, production, and quality performance are simultaneously available to all those individuals who are required to make operational decisions, whether that is a machine operator or the vice president of operations. It will provide the architecture for interfacing all CAE, CAD, CAM, and resource systems (i.e., parts, materials, workforce, schedules, and configuration definition).

Statistical process control and other quality tools. Statistical data can be used to understand what features or characteristics should be measured and analyzed to simplify, improve, or more effectively control the process, product, or service. Statistical data provides an able tool for problem focus and solution when correctly analyzed and used by trained staff.

Implementing SPC.   Once management, supervisors, and operators are trained in the concepts of SPC fundamentals, which includes control charts, process capability study, design of experiment, statistical control limits, Pareto charting, and cause-and-effect analysis, concurrent engineering teams (working groups) are formed to develop and review manufacturing process plans that establish the primary production sequence in process steps and also specify the production parameters to be controlled by SPC. The teams are specifically charged with developing and implementing SPC into the production operations.

SPC should be implemented in parallel with the manufacturing efforts in place and integrated in a noninterference manner, if possible or practical, in a workstation-by-workstation progression. Once integrated into a workstation, the system at that station is debugged to assure minimal impact on the performance of operator(s), while demonstrating the improvements in yield that can be directly attributed to SPC techniques. The objectives at each workstation are to

1. Provide statistical data for defect prevention, not necessarily defect detection

2. Reduce quality inspection

3. Give the operator the ownership for making and accepting defect-free products

4. Give the operator simple, basic ground rules for stopping the process

First steps.   Determine the feasibility for implementing SPC at each workstation. Each step in the process at a workstation is reviewed to determine its quality characteristics. A team composed of SPC-trained individuals directly involved with the performance of that particular workstation (operator, supervisor, manufacturing engineer, and quality engineer) conduct the review and decide whether SPC can be used to provide real-time control of product quality, reduce nonconformances, and improve productivity of the process.

A capability study is performed as part of the review, providing attributes and variables analysis of the machines and the subprocesses involved. Pareto analysis of defect histories, including scrap, rework, and associated paperwork and a feasibility study to determine potential quality improvements, cost reductions, or production enhancements, is also included in the review. If the results of these studies indicate that quality or economic value can be added to the workstation by implementing SPC, the study will be submitted to higher authorities for approval and, when appropriate, submitted to the customer.

Following approval and implementation, daily, then weekly, meetings of the implementation team are held to ensure that the candidate SPC installation and performance is running as expected or to determine whether further adjustments are to be made. As workstations demonstrate, through an arbitrary trial period, that SPC operations are successful, then that SPC portion of the TQM system, with its operator, can be certified for product acceptance.

Depending on the size of the factory, the implementation process could take several years before the last workstation is certified. Attempting to address all production workstations simultaneously is beyond the capability of even the largest companies and, if attempted, could result in counterproductive results. Instead, workstations should be separated by priority listing and pursued serially with the number 1 priority receiving the first attention.

The top priority workstations should be those that appear to be a major source of variations, are near the start of the production line, and that would likely require the least amount of span time or capital expenditure to correct.

Follow-through. After SPC certification has been achieved and SPC data is used to accept products, quality control will monitor the data collection system to assure compliance. In addition, a minimum of one part per shift per active certified workstation will be remeasured. Quality control will also verify that all data points are within the control limits or that documented cause and corrective action is integrated and verified for points out of the control limits.

Certification criteria. Review a minimum of 30 data points for each characteristic in each workstation process.

1. The capability ratio must be less than or equal to 75 percent.
2. Capability index, i.e., the specification width to the process width, must be greater than, or equal to, 1.33.
3. Data points must be within control limits.
4. If more than one data point out of 35 (depending on quality improvement objectives) is outside the control limits, the process is not in control and corrective active will be taken.

Supplier flow-down. Implementation of SPC flow-down requirements to the suppliers is an ongoing process supported by quality control and procurement organizations. Every process in the supplier's factory that influences the product is a candidate for application of SPC techniques to help reduce, eliminate, and prevent nonconformances. As supplier data is developed, analyzed, and interpreted to show that SPC is achievable, a plan to reduce receiving inspection requirements can be initiated. If the plan is approved, product acceptance from the supplier, based on SPC data, can be incorporated into the TQM system.

Natural tolerances. Each process has its own natural tolerances as determined by the machines, chemicals, materials, operators, and environments. The sum of these tolerances uniquely determine the capability limits of the process. If the sum of these tolerances renders the process capability less than the specification, then tighter controls and/or process modifications are needed so that the natural spread of tolerances can be narrowed to meet the specification with a comfortable margin that produces a capability index of 1.33. If process modifications cannot correct the situation, then design modifications may be needed.

General areas of vendor deficiency. Control and use of measuring and test equipment by the supplier should be in compliance with MIL-STD-45662. Supplier measuring and test equipment used to buy off final products are required to be calibrated at established intervals. Products measured or tested by equipment or devices with outdated calibration cannot be trusted. Outdated calibration is a persistent problem for all companies and requires special attention which can be facilitated by automatic flagging and follow-through. In the event final inspection has been performed by equipment with expired calibration and shipped to the customer, the customer should be notified and the equipment recalibrated and certified. If the equipment remained accurate as verified by recalibration and did not require adjustment, then reinspection of the shipped lot would not be necessary. If equipment adjustments or rework were needed, then the lot would need to be returned to the supplier for reinspection with properly calibrated equipment.

Process control also implies that any instrument monitoring a process function or activity should be calibrated. The selection of instruments for process application should be done with ease of calibration as one of the prime requisites.

Another area that is persistently neglected in addition to calibration is the maintenance of a system of identifying the inspection status for perishable items, such as dry-film photoresists, solder pastes, and adhesives. Identification should occur at receiving with limited-life labels, with expiration dates, attached. Physical audits by quality control of all materials is often necessary at periodic intervals.

Inspection status is another problem within stockrooms for traceability of matériel pedigree. When material sheets are cut or component containers split to fill a job order, the original label or accompanying paperwork containing purchase order (PO) numbers, dates, part numbers, lot numbers, and quality acceptance are separated and often the remaining stock is left without a traceable pedigree. All paperwork should remain with the leftover matériel and job folders stamped by quality control for the issued material.

When auditing a supplier, the following points are key to a thorough quality inspection system.

Examples of process variables.   The following items are a few examples of critical features taken randomly from throughout the SMT production sphere that can seriously bias process results beyond their natural tolerance limits: oxidation levels on component leads and PWB pads, component lead coplanarity, solder-to-flux ratio of paste, adhesive registration, solder screen squeegee blade pressure, heater temperature controls, conveyor speed, deionized water purity, and machine adjustments made by each shift operator.

SPC versus quality levels.   Conventional SPC can be used as the sole quality tool in achieving $\pm3\sigma$-level quality. But as the need for higher quality level continues to rise, additional measures are needed. SPC is necessary but not sufficient at the higher quality levels. Products need to be designed with a robust imperviousness to process variations, and processes must be developed with more narrow control limits. Quality goals for some world class companies are already at the $\pm6\sigma$-level and going higher.

Design for manufacturability (DFM) is an absolutely necessary prerequisite to achieving 4σ-level quality or higher, and design for robustness (DFR) against vendor material parametric variations is another key ingredient to achieving quality levels above 4σ. DFM and DFR are engineering-based activities for improving quality beyond the customary levels. Manufacturing can introduce Shigeo Shingo's (Japanese consultant) methods of mistake-proofing the process by installing simple, inexpensive mistake detectors *to catch potential errors before they happen* and therefore drive error-detection time to zero. Mr. Shingo's views on SPC, however, are questionable.

Computer integrated manufacturing (CIM) as a vital part of TQM.   Companies often embark on a number of independent data-based automation initiatives that grow and progress without coordination and consideration for integration. Companies have tended to enter into these computer-based manufacturing subsystems without having a strategic overall plan. What is needed, many times, is an organized program to bring together these computer-based projects to develop, integrate, and implement a CIM system.

These early initiatives into automation of the factory would likely include the following:

1. *Inventory control.*   This program consisted of bill of material, material requirements planning, inventory control, purchasing and receiving interfaces, and finance interface.

2. *Manufacturing engineering.*   Generation and maintenance of the manufacturing bill of materials, routings, standards, process plane, tool drawings, and tool fabrication plans.

3. *Factory control.*   Work order priorities, factory work control, and labor data collection.

4. *Performance work measurements.*

By integrating these earlier subsystems and acquiring new ones as needed, an on-line manufacturing resource planning system, interfaced with on-line purchasing, production planning, and distributed factory control, can be put in place. Such a system would be flexible, in that the independent subsystem or modules would permit ease of configuring the overall system to achieve a comprehensive array of interactive on-line displays and collection terminals. Financial data would be compatible with operational data, and current costs could be tracked in real time. With the inclusion of bar codes, a paperless factory could become possible. Accurate, up-to-date purchasing data could time-phase supplier lead times. Real-time supplier status on all open purchase arrangements could be available for JIT (just-in-time) systems as well as non-JIT systems. Performance measurements on buyers and suppliers would be automatically available.

A comprehensive list of potential subsystems to be added to the preceding lists as part of the CIM system and consequently TQM might include

1. Engineering bill of material

2. CAD detail and assembly drawings and database artwork for PWBs

3. CAE product schematic netlists could be automatically made available for test tool design and planning

4. CAM

5. Master production schedules

6. Capacity requirements planning

7. Finance

8. Plant performance

9. Configuration tracking

10. Quality

11. Payroll

Another benefit from the CIM system is an increased real-time inventory accuracy that approaches 100 percent. Coupled with an automated storage and retrieval system of carousels, conveyors, and bar codes, the inventory control could approach SPC failure-free operations with lower inventory levels.

Technical audits as part of TQM. Technical audits gather factual information and access compliance to contractual, regulatory, and company policies and standards procedures in hardware compliance, software compliance, systems adequacy, internal functional compliance, and housekeeping. When deficiencies are found in any audit, the root causes should be determined and immediate corrective action taken.

Performance measurement. A real-time performance-measurement-system capability at the production working group level and a system for accountability and ownership of performance at the first-line supervisor level can be valuable for the workers, the company, and TQM.

A team composed of all the hourly workers in each manufacturing work center, the area supervisor (the team leader), and the area general foreperson, along with the manufacturing engineer and the quality personnel assigned to support the area is formed to provide focus and work-center involvement in quality performance improvements and performance measurement system. The system would provide regular performance measurements on factory yields, scrap, schedules, standards, overtime, and lost time at the work center. Goals are established for each measurement, and all data are prominently displayed in each work center. Goal settings, real-time performance feedback, ownership and accountability for results, public display of results, and prompt recognition for achievements would help motivate all those individuals involved. Commitment and involvement of mid- and top-level management is key to the success of such a program.

Outstanding performance should be recognized in an outstanding way. Employees who best symbolize TQM should be recognized with immediate monetary awards coupled with an invitation to worker and spouse to an annual top-employees posh banquet.

TQM in the office.   TQM includes the quality of performance throughout the company. This includes the office work as well as the factory and field service work. Since almost all work is done in accordance with a process of one form or another, whether formal or informal, SPC can be applied to many of the office routines to improve the quality level. Promotion of error-free, single-pass production should be as much an objective in the office as in the factory. Automation can also be as much a positive attribute toward quality improvement in the office environment as in the factory environment.

Customer satisfaction.   A clear focus to satisfy all customer requirements at competitive costs is as valid in the white-collar world as in the blue-collar world, from the initial customer contact through delivery of the final production order and into the follow-through field support service. TQM, as it relates to customer satisfaction, is all-inclusive from conception through the entire life cycle of the product.

Technical operations.   Technical operations should embody the principles and elements of TQM in all phases of engineering design, development, production, and field support. Concurrent engineering should become a way of life for the engineering department. With concurrent engineering, error-free, single-pass, excellent designs become much more feasible and achievable. CAE and CAD facilitates quality improvements in the design processes at lower costs.

All office workers.   All office workers can focus on information computing systems, simplifying processes, and streamlining the interfacing between functional groups. Create a culture committed to continued improvement to enhance productivity and cost-competitiveness through performance measurement technique and team efforts. Continue to consolidate administrative functions wherever possible and implement SPC to monitor the consolidated functions. Be very diligent about continuously improving the quality of communications with the customer.

Random observations.   Some random observations for SPC in the office arena are listed here:

1. Start with training of SPC fundamentals for all office workers and supervisors likely to be involved with SPC.
2. Construct a methodology to identify opportunities for improvement.
3. Identify chronic waste of material, time, and other resources.

4. Have every person identify his or her personal customer(s) and learn the customer's needs and wants.

5. Motivate workforce to satisfy the customer, with strong emphasis on satisfying external customers but with equal vigor in satisfying internal customers as well.

6. Add value to every effort—make a difference daily.

7. The heroes of the future will be those individuals who make TQM and SPC successful and profitable for the office environment.

8. It is in the best interest of each individual office worker to get involved with TQM and SPC with intensity and long-term commitment.

9. Work toward making office work processes robust and relatively immune to process variations.

10. Develop a long-term mentality backed with detailed, realistic plans that have improvement responsibilities assigned and timetables established.

Top supplier recognition.   A periodic conference needs to be held bringing all the top suppliers together for a one-day session discussing supplier issues, business prospects, and quality issues. Top suppliers could receive cross training in SPC, share lessons learned, and become recognized as a preferred supplier and perhaps enter into an alliance. A procurement rating system could be initiated on the basis of historic data of the parts received versus the parts accepted, dates met, and competitive price analysis derived to support awards to overall best bidders or suppliers. Award more business to proven performers.

Summary.   TQM as presented here is not meant for every company. Very few could afford it or have the staff and scope of operations that make such a system necessary. However, adapting the theme of TQM is very necessary for every company. Each company would then work out the details of quality improvement as best fits their circumstances.

B

Procurement Requirements for SMT

Procurement of SMT components is much more stringent than it is for IMT components. Although SMT components are electrically the same as IMT components, the acceptance quality levels (AQLs) are higher. With three and four times as many SMT components on an assembly, each SMT component has to have a higher AQL to maintain the same assembly yields as the IMT boards. Mechanically, SMT components need higher quality levels because of the automatic placement requirements that use the component body for placement. Also, SMT components are subjected to greater stresses during assembly processes.

Solderability. Solderability and leaching resistance testing are required as part of the supplier's preship test or user's incoming inspection. Unlike IMT, in which lower-quality solderability can be overcome for commercial applications at assembly by using more active flux, SMT depends on high-quality solderability and the less active flux. Because of finer lines and more dense component packaging, SMT assemblies are particularly sensitive to ionic contamination, which is more likely to occur by using flux with increased activity. A higher degree of solderability of the components is required to prevent tombstoning and skating of small, discrete devices. Additionally, the PWB must also be tested for solderability.

Coating the solder pads with solder does not protect them over a longer period of time (over 6 months for some storage environments and 1 year for environmentally controlled storage). Electroplating solder on PWBs produces substantial variations in the solder composition across the PWB. Hot-dipping eutectic tin-lead to a minimum of 0.018-

mm thickness is required to pass steam aging tests. This is the minimum thickness that will retard the solder pad surface from developing intermetallics.

High-temperature resistance. Unlike IMT components which are somewhat thermally protected from the wave-soldering temperatures by the PWB thermal insulation properties, SMT components are thoroughly drenched in infrared or vapor-phase reflow solder temperatures. SMT components, therefore, must be tested to higher thermal resistance levels to ensure that physical deterioration does not occur during the soldering process. SMT components must pass a direct immersion in a liquid or vapor at 215 ± 5°C for a minimum of sixty seconds (60 s) *and* must also be able to pass a direct immersion in liquid *only* at 260 ± 5°C for a minimum of ten seconds (10 s). The same component is not required to pass both tests, but samples from the same lot are required to pass both tests. Components should be preheated to within 100°C of the test temperature for one minute (1 min) prior to immersion. Preheating temperature rise should not exceed the normal soldering rates.

Solvent resistance. Existing solvent testing (method 215 of MIL-STD-202 or method 2015 of MIL-STD-883) are geared to cold cleaning methods. However, solvents used for the cleaning methods employed for company processes should be used to test components under similar cleaning environments (jet spray, heated vapor, immersion, etc.).

Visual inspection. Visual inspection of SMT components and PWBs is beyond the acuity of the unaided eye to detect faults. Visual inspection must be done under magnification.

Component packaging. There are four general methods of packaging SMT components for transportation, storage, and assembly handling:

1. Tape and reel
2. Magazine (or stick)
3. Bulk
4. Waffle pack (or matrix trays)

Tape and reel is used for large production and automation. Taped components are individually packaged on 8-, 12-, 16-, 24-, 32-, 44-, and 56-mm tape widths. Embossed plastic tapes are preferred over paper tapes on automatic placement systems. These tapes, in the 8-mm size, can be stuffed with 3000 to 4000 chip components that are a maximum of 8 mm thick. Paper tapes shed paper fibers, clogging the vacuum pickup heads. Paper tapes, in the smallest size, can carry up to 2000 chips having a maximum thickness of 1.1 mm. One of the major problems with tape packages is that the components are sealed, making inspection and testing of taped components by the user impractical.

Bulk packaging is also used for large-scale production and automation. The packaging is comparatively inexpensive, with a compact volume, and components are readily accessible for inspection and test. Components are, however, easier to be damaged in bulk packaging and difficult to orient properly.

Magazines (sometimes referred to as "sticks," "rails," or "cartridges") are the most useful package for short-run automation. They are easy to use, but because they are expensive, have a small capacity, and because of their tendency for components to separate, they are not used for larger-scale production.

Waffle packs, or matrix trays, are used for prototypes, short production runs, or large components with fine-pitched leads.

Case Study—Super-PWBs for Next-Generation Supercomputers

Supercomputer Systems Incorporated, a new startup company funded by IBM, is now building the next-generation supercomputer using 78-layer organic PWBs that measure 151 × 151 mm and are routed with conductor line widths of 63 μm. These boards boast a propagation delay of 58 p s/cm with a characteristic impedance of 57 Ω and a crosstalk of less than 3 mV/cm.

Using basically conventional packaging techniques, fabrication of these boards has been made possible with the choice of dielectric material, the use of sequential lamination, and enhanced optical registration.

Dielectric material. Rogers Corporation's RO2800 randomly reinforced, spherical shaped, ceramic-filled PTFE advanced material is used because of its smooth surface, ease of laser-drilling very small holes, 2.8 dielectric constant, dimensional stability, and a dielectric CTE that matches the CTE of copper to reduce or eliminate stress between the PWB base material and the small lines and vias.

Sequential lamination. Boards are built in stages using subassembled cores. Subassembled cores generally consist of 12-layer MLBs that have been fabricated using all the normal processes, except for a special class 10 clean-room lithographic machine. Normal bare-board testing and automatic optical inspection of each layer and the final laminate are used for the final core subassembly. Generous use of fiducial marks is made at the individual layer level to take full advantage of optically assisted registration techniques at each step of lamination and drilling.

Because of the absence of embedded glass reinforcement fabric, the material and core subassemblies are conducive to laser drilling of the 75-μm-diameter buried vias and the 150-μm-diameter blind and through-hole vias. The exceptional dimensional stability of the selected material has made registration and drilling of small holes possible.

Faulty layers or core subassemblies can be repaired by either cutting short circuits, using a xenon laser, or welding gold ribbons across spaces in any of the lines. Core yields, for prototype production, are nearly 90 percent, subsections are approximately 80 percent, and final PWBs are about 60 percent yields. These are good yields for such an advanced, complex PWB.

JEDEC Die-to-TAB Standardization Recommendations

JEDEC has proposed the following eight points as the basis for a Die-to-TAB interface standard:

1. Use an odd number of I/O pads on each side of the die with the odd pad being in the geometric center of the die.
2. Orient the centerline of the die with the centerline of the TAB frame.
3. The center test pad at the OLB should be on the centerline of the TAB frame.
4. Die pad pitch should be identical on all four sides.
5. Die pad size should be at least $60 \times 60\ \mu\text{m}$.
6. Die pad pitch should be at least 100 μm.
7. Die pad center to die edge should be at least 120 μm.
8. Die corner pad clearance should be at least 280 μm.

A ninth point that is critical and should be considered in addition to the eight points recommended by JEDEC is that each I/O-size die (20 pins, 28 pins, 32 pins, etc.) should have its own fixed pattern, regardless of electrical function.

Index

Active devices:
  categories of, 61–62
  rework of, 220, (figure) 221
Acrylic elastomers, 19
Adaptive leads (add-ons):
  for LLCCs, 63–64
  rework of, 218–220
Adhesives application:
  and component retention, 117
  dispenser for, 118–122
    dot sizes, 118
    high volume, 120, (figure) 121
    low volume, 119, (figure) 122
  and double-sided CCA, 40–41
  and LLCC standoff, 129–130
  repair, 217, 223
  screen/stencil, 120
  soldering, 122
  and thermal conductive film, 55
Alumina ceramic, 232
Aluminum nitride, 232
Aqueous cleaning, 175
Audits, TQM, 256
Automation, total, 7
Azeotropic solvent, 174

Bare-board testing, 204
Barrel-cracking, PTH, 38–42
  and PWB materials, 38–39
  and strain, 39–42, (figure) 38
Bathtub curve, 41–42, (figure) 43
Bed-of-nails, 205, 210
Benefits of SMT, 3–5
  size and weight (table), 4
  two essential features of, 3
Benzocyclobutene (BCB), 231
Brute force contrivance, 44

Bulk packaging:
  uses of, (App), 269
  form of, 69
Butt lead (I-lead):
  an add-on, 64, 216–217
  shape of, 63
  soldering standards for, 172, (figure) 173

CAD/CAE in SMT, 26–27
  manual layout of, 26
Carpets, for ESD, 196
CCAs, cooling, 24
Ceramic packaging, 7
Certification of SPC, 259
CFC solvent elimination, 175–177
CFC substitutes:
  acqueous/semiadcqueous cleaning, 177
  alternate solvents, 176
  nitrogen benefits of, 140–141
  no-clean flux, 176
  terpenes, 176
  water-soluble flux, 177
Chip mounted technology (CMT):
  and cooling, 24
  description of, 236
  material for, 51
  products, 7
Chip-on-board (COB) replaced, 236
Circuit card assembly (CCA), 4
Cleaning, postsoldering:
  agents for, 125
  assemblies for, 123–124
  cleanliness measurement of, 177–178
  and flux residue, 155–156
  and heat, 125
  immersion, 125

ABOUT THE AUTHOR

Frank Classon has over 30 years' experience in airborne, mobile, and stationary electronic packaging as a design engineer, program manager, and department head. He was responsible for introducing SMT on production products as early as 1978, and has been actively involved in SMT design, development, and manufacturing ever since. He was recently General Chairman of the IPC Electronic Packaging Design Committee and a teacher of SMT at Valencia College and Technical Seminars, Inc.